WAR

PAST, PRESENT & FUTURE

Jeremy Black

SUTTON PUBLISHING

First published in 2000 by
Sutton Publishing Limited · Phoenix Mill
Thrupp · Stroud · Gloucestershire · GL5 2BU

British Library Cataloguing in Publication Data
A catalogue record for this book is available from the British
Library

ISBN 0-7509-2394-6

Typeset in 11/14.5 pt Sabon.
Typesetting and origination by
Sutton Publishing Limited.
Printed and bound in England
by J.H. Haynes & Co. Ltd, Sparkford.

For Spencer Tucker

Contents

Preface

Abold, not to say presumptuous, title requires an explanation, especially as this is the preface to a long essay rather than a multi-volume work. I have written at some length about particular aspects and periods of military history, but have not hitherto attempted to develop ideas at any length. The title of this book reflects my view that in order to understand present circumstances and future possibilities it is necessary to appreciate military history; and, furthermore, that a consideration of these very circumstances and possibilities throws light on that history. If this is one of the themes of this book, it interacts with others, including the role of cultural suppositions in conflict, the need to consider force within as well as between states, the failures of Eurocentric assumptions, and the weakness of a chronology and causation focused on technology, specifically on military revolutions defined by new technologies. Particular attention is devoted to civil conflict because it is seen as overly neglected in military history.

Furthermore, civil conflict highlights the role of contingency and thus opens the way to the counterfactual speculations that are important when assessing the consequences of warfare. This counterfactualism is intended to contest the determinism that affects some of the 'new military history' which has moved away from operational history with its kaleidoscope of possibilities. The stress on contingency is linked in this book to the issue of the rise of the

West, which is a central question in military history. 'Revolutions in military affairs', to the extent that they exist, are shown to be essentially culture-specific, and military cultures are presented as both more flexible and more mutable than is generally allowed. This is related to the degree to which societies and states can come to accept a quite wide variety of circumstances as 'normal'.

I have benefited greatly in developing these ideas from over two decades of teaching military history, first at Durham and then Exeter. In addition, I have profited from the opportunity to give lectures to outside audiences and to take part in conferences. In the period in which this book was written the latter included lectures to the 1997 Anglo-American Conference, the 1999 conference of the Consortium on Revolutionary Europe, the 1999 University of Virginia programme on the American Civil War, the millennium series of the Birmingham Historical Association, and lectures at Adelphi University, the University of Illinois, Urbana, the University of North Carolina, Greensboro, the College of William and Mary, MIT, Peterhouse College, Cambridge, Elizabeth College, Guernsey, and Sherborne School. I am most grateful to those who invited me. I would also like to thank Rob Johnson, Harald Kleinschmidt, Gervase Phillips, Dennis Showalter and Spence Tucker for commenting on earlier drafts of parts of this book, and Christopher Feeney and Helen Gray for seeing it through the works at Sutton Publishing with exemplary speed and efficiency. It is a great pleasure to dedicate this book to a good friend and fellow military historian.

Abbreviations

Add. Additional Manuscripts
AE CP Paris, Archives du Ministère des Relations Extérieures,
     Correspondance Politique
BL   London, British Library
Bod. Oxford, Bodleian Library
JMH   *Journal of Military History*
PRO   London, Public Record Office
RAM   Revolution in Attitudes to the Military, see p. 245
RMA   Revolution in Military Affairs, see pp. 69, 239
SP   State Papers
WO   War Office

Unless otherwise stated, all books cited are published in London.

ONE

Introduction

Military history has not played a major role in the academic community for decades, and scholarly work on modern forces and warfare is also limited. The situation varies in particular countries, but the overall picture is bleak and has attracted comment.[1] This is unlikely to change, for hostility to military studies is, simultaneously, seen to betoken opposition to militarism and war, and also to reflect the rejection of the structures, practices and ideologies of authority that characterizes so much of the self-image of the academic community. This is particularly so in North America and Europe, where academics are numerous, relatively well funded, and apt to see their own views as globally applicable. A view that war is an epi-phenomenon (an episode lacking fundamental significance) and that it is less important to human history than structural characteristics centred on socio-economic, environmental or geopolitical considerations characterizes most scholarly work that adopts a structural perspective, whether or not it is Marxisante in type. In short, war can be left outside the historical mainstream.

This relative neglect of the military dimension is paradoxical for a number of reasons. One is intellectual method and fashion. The dominant intellectual agenda of the Western academic community is based on force. Knowledge is seen as a power system: it is about struggle and is, itself, the product of struggle. The relation of knowledge to power has indeed encouraged a militarization of

1

analysis and discourse: the language used encompasses ideas of hegemony, spheres of contention, and boundaries. Furthermore, modern and historical society are seen in these terms: communities – political and social, household and larger – are presented as created, structured and changed by power. Coercion, not consensus, is the theme. Conflict and control are seen as the norm in relationships.

In the case of the household, it is indeed possible to emphasize power and coercion without bringing in the military, but this makes far less sense at the level of societies, states, and international systems. As a consequence, discussion of the nature and role of military force plays a role in theories of state and international development, and this, indeed, became more the case in the 1990s. What has been described as the attempt to bring 'the state back in'[2] to grand accounts of political development led to a focus on the consequences of war. This is especially true of its impact on governmental expenditure, administrative development, practices of political consent, and social structures.[3]

It is ironic that such analysis is not generally produced by military historians. They of course do not 'own' the subject of the history of war and its consequences and, indeed, tend to be only interested in military consequences. Furthermore, there are various ways in which the subject can be profitably approached. Yet it is important that the insights of the military historian are offered. The development of states and societies, and the creation and operation of global power systems are two major questions that concern historians and other commentators at the start of the new millennium. Military historians should seek an input to both, and indeed they are interrelated. Both questions can be approached in the knowledge that there is a measure of convergence between military history, in so far as it can be reified, and other aspects of the subject. In particular, the recent emphasis on the culture of military organization and practice, as a source of explanation for operational characteristics and success,[4] bridges traditional military and non-military concerns and methods.

Work on the first question, the role of war and the military in the development of states and societies, is most common in the case of the situation in Europe, but is far from restricted to this continent. Nevertheless, assumptions and models derived from the European situation are widely applied elsewhere; they provide the very vocabulary of the subject. Such work also constitutes the 'sub-text' of much military history, specifically of the established chronologies and causations of military change. Thus, the notion of an early-modern European Military Revolution was originally advanced by Michael Roberts as part of an explanation of state-building.[5] Similarly, changes in the character of European warfare in the 1790s have been related to the impact of the French Revolution, and developments thereafter have been related to the impact of industrialized society and of nation states, able and determined to mobilize their citizenry.

Military historians have made a distinguished contribution to the study of the interaction of war, armies and state development in the European context, although — as is characteristic, in general, of their methodology — this contribution has tended to be empirical rather than theoretical.[6] Such empiricism is very valuable. For example, to take one particular issue, recruitment is best considered in this way. Theoretical reflections on the relationship between the military and the state, such as on the impact of revolutionary enthusiasm, can be greatly refined by such studies. This is also true of the attempt to assess the social configuration of the military by considering the extent and consequences of aristocratic officership.[7]

Such questions have much to do with armies and navies as means to wage war, not least because their socio-political context is instrumental to the motives and means by which forces are sustained in war. Far from there being simply one dimension to the operational aspect of war, it was as much a matter of the provision of pikes as the push of pike on the battlefield. Furthermore, issues such as recruitment and officership were important to cohesion and morale, and thus to effectiveness in battle conditions. Indeed, the more scholars, as they have done over the last three decades, emphasize cohesion and morale as factors in success, whether on the battlefield or maintaining a rate of march or avoiding insanitary practices, the more it is important to probe these socio-political aspects.

3

This is true both of symmetrical and of asymmetrical confrontation and conflict. In the former case, where there is a similarity in weapons, weapon systems, organization and tactics, then it is necessary to probe other causes of relative capability and success, as well as the 'variables' of conflict, such as weather and terrain. In the latter case, where there is no such similarity, it is important to consider how differences in military characteristics were understood, confronted, employed and, sometimes, overcome.

Such asymmetry is commonly understood in terms of clashes with other and different societies, but it was, in fact, present within most states, for political opposition was often expressed, and more frequently suppressed, through force; and the same was true of social dissidence, in so far as they were separable. In general, conflicts within states have attracted insufficient attention from military historians, and are frequently neglected when considering the future of both the military and warfare. The first remark may appear risible, given the quantity and quality of work on the American Civil War, but that struggle, like the English Civil War, focuses the issue, because civil conflict tends to attract attention largely when it approximates to international wars, in short when there were clear-cut armies, command structures, battles, and areas of control.

That, however, was not typical of civil confrontation and conflict, and the asymmetry of much of this – for example, in the eighteenth century, of Andean peasant risings[8] or rebellions by both Chinese and minority peoples within the Chinese empire – is worthy of consideration. Such challenges could be as important as international struggles as a cause of commitment for armies (although not navies), and in some cases even more so. Thus, the San-fen Rebellion, the War of the Three Feudatories, in China in 1674–81 was a bigger challenge to the Chinese government than confrontation and conflict with Russia in the Amur valley in 1685–9. The White Lotus Rebellion in Shaanxi in 1796–1805 was a more serious challenge than the conflicts in Tongking in 1789 or with Nepal in 1792. Similarly, the Taiping Rebellion of 1850–64 was a more serious challenge than the Opium War with Britain of 1839–42, although the latter was far less peripheral than the

4

conflicts in the Amur Valley, Tongking and Nepal. Chiang Kaishek was more concerned about warlord and Communist opposition in the early 1930s than about Japanese gains in north-east China, although many did not share his priorities.

In areas where international conflict was uncommon, such as Latin America after 1941, the prime activity of the military could be asymmetrical internal confrontations, and this was important for the development of military cultures. Furthermore, more generally, it is necessary to overcome the relative neglect of the domestic use of force in theories of military organization and development. The importance of this usage in assessing the relative military capability of states needs to be stressed, and this is best done by locating the military and war in their socio-political contexts.

Asymmetrical conflict has attracted most attention as an aspect of the world question: the issue of global hegemony or, at least, of dominance in a major part thereof. This issue is commonly approached from a Eurocentric perspective and becomes a discussion of how and why the Western European maritime powers came to control, directly or indirectly, so much of the world by 1914. This question has been taken on into the present day, and towards the future, by considering the position of the USA. Although it is indeed the leading Western state, the nature of American power, however, is different to that of the Western maritime states. The major difference is the detachment of colonialism from imperialism by the Americans, whose public culture is that of a state with a strongly democratic ethos: after the conquest of the Philippines in the 1900s, the Americans abandoned colonialism, at least in so far as formal control was concerned. As a consequence, American imperialism or global-range dominance has generally been acceptable to its public opinion.

Before discussing the rise of military dominance, it is worth suggesting that insufficient attention has been given to other ways to create and structure links and 'control'. Secondly, in so far as the nature and use of force are the key questions, it is possible that more attention should be devoted to other powers, both those elsewhere in Europe and, more significantly, outside the Continent.

The first point may appear paradoxical in a book urging greater attention for military history, but, whereas generally the view of the military historian is insufficiently heeded on the development of states and social systems, there has frequently been an excessive concern, in consideration of the expansion of Europe, with force at the point of contact. This might seem to be the obvious subject of discussion, but it needs to be complemented by, and understood in the context of, consideration of other aspects of European contact with non-Europeans and other military purposes. If powers are assumed to be on an inevitable course of territorial domination and aggrandizement, then the military aspects of conquest play a greater role than if, for example, the stress is on how the possibility of the use of force enhanced commercial opportunities.

Trade was indeed one of the major ways in which links were created and then maintained. Trade is not separate from force, as the concept of protection cost developed by Frederick Lane made clear,[9] or to probe another angle, as the relationship between agrarian empires, such as Rome and China, and 'barbarian' neighbours demonstrated;[10] but trade is not coterminous with conquest. The relationship between trade and the display of strength, and the importance of warships in shows of force, were captured by Captain William Cornwallis in his report to the British Admiralty about a voyage to the River Gambia in West Africa in February 1775:

> Upon sending an officer up to James' Fort, I was informed by the commanding officer that the French had spirited the natives up against the English, and that he had been obliged to take a schooner of some force belonging to the traders into the service in order to supply himself with water; I thought the appearance of a man of war might be of service. I therefore went up the river in the *Pallas* to James' Fort, which I found in great distress for want of stores, and particularly gun-carriages, not having above three or four serviceable ones in the Fort, and most of their guns rendered totally useless for want of them . . . I stayed in the River eight days, during which time we got the king of the country on

board and showed him all the civility we could; he seemed very well pleased, so I hope all will go on well again.[11]

Western scholarship has been overly apt to identify trade with naval power, and conquest with armies, but such a distinction is inappropriate. As far as Europe is concerned such an analysis arises in part from a misleading Whiggish, beneficent account of British and, to a certain extent, Dutch maritime strength. More generally, the role of trade in landward relations between non-European societies is frequently unduly minimized.

The nature of links is important because it relates to the motives that underlie relations between different states, societies and peoples. Clashes and conflict can best be understood in terms of these motives, and thus they are crucial to the issue of the purposes for which military forces are designed. Instead of the so-called realist model of international relations, with its thesis of an unstable, anarchic international situation, and of states carefully calculating advantage in war and the creation of deterrent alliance systems, it is more appropriate to consider the contingent role of concepts, specifically the impact of what have been termed strategic cultures. These affected both the choice of opponent and the purpose and nature of warmaking.[12] There is little point in discussing why one side won a war unless the question 'what did victory mean' is addressed. Looked at differently, military capability is a matter of fitness for purpose, but this purpose is culturally constructed.

This point emerges clearly both from consideration of the frequency of conflict within a core international system, as opposed to war with powers in a recognizably different system, and from studies of military Westernization over the last quarter-millennium.[13] In the first case, it is necessary to note the contingent character of warfare, specifically the cultural factors and decision-making processes that influenced the choice of whether to fight powerful neighbouring powers or to turn to more distant targets. This was not simply the situation with nineteenth-century European imperialism, with competition, for example, for territorial expansion in Africa, South Asia and Oceania, rather than conflict in Europe in the last quarter of the century. Earlier, in 1761, a British diplomat had noted: 'It is observable that the Russian sovereigns, instead of taking the fairest opportunity, during the troubles of Persia, to erect a mighty, Asiatic empire, have turned their views wholly upon Europe.'[14]

In military Westernization, adoption of weaponry, and even tactics, is of limited value unless there are accompanying cultural changes. These changes have to occur both within the military – whether drill and discipline, or meritocratic promotion or the creation, in modern specialized forces, of a bureaucratic-technical ethos – and in society as a whole. The latter is especially, but not only, the case with conscript armies. Yet such changes may be actively resisted, both within the military and in society as a whole, because there is unwillingness to revise the cultural understanding of the purpose and use of force, military authority, and the nature of merit. This was true, for example, of the Ottoman empire in the late eighteenth and early nineteenth centuries. The overthrow of Sultan Selim III in 1807, when he sought to reform the janizary auxiliaries, the dissolution of Selim's new model force, the *Nizam-i Cedid*, and the failure of Mahmud II to re-establish control over the janizaries in the 1810s all indicated the deeply-rooted ideological, political and social obstacles to Turkish military reform. This also suggests the limited value of considering such reform in isolation from wider social and cultural developments, an important point with reference to the modern and future military.

An understanding of variations in the nature and use of force, and their validity in particular socio-cultural contexts, necessarily directs attention away from war primarily understood as the clash of different weapons and weapon systems. The latter offers an apparent basis of global judgement of capability and effectiveness, but this is of limited value outside an understanding of cultural suppositions. The situation emerged clearly when peoples were 'conquered' but then refused to surrender, in short when the defeat of field forces or the capture of major centres did not lead to an acceptance of defeat. Thus, in South-East Asia in the fifteenth and sixteenth centuries, the notion of fighting for a city was not well established culturally. Instead, the local culture of war was generally that of the abandonment of cities in the face of stronger attackers, who then pillaged them before leaving. Captives, not territory, were the general objective of operations. European interest in annexation and the consolidation of position by fortification, however, reflected a different culture. Similarly, the Iroquois in North America were concerned with glory, honour and revenge, rather than economic issues. In particular, it is necessary not to extrapolate their interest in control over the fur trade into a set of motives for policy similar to those of Europeans.[15] This is not simply a contrast restricted to a period of time. Indeed, different assumptions of what control over territory and victory in conflict mean play a role in modern conflicts, and will doubtless also do so in the future.

In South and South-East Asia, European annexation in the early modern period was designed not to lead to an extensive territorial presence, but to provide safe ports and entrepôts (markets), such as Malacca, captured by the Portuguese in 1511. Conflict, for the Europeans, was a matter of securing their position in trading networks, specifically in the coastal zone of transhipment. Rather than thinking of contact in terms of conquest, it is important to note that that was a contingent, not an inevitable, consequence of the European presence. At one, admittedly untypical, extreme was the attitude of Revolutionary France, as reflected in the report that Journu-Auber, on behalf of the Colonial Committee of the National Assembly, made to the Assembly on 7 January 1792 about what French policy should be towards Madagascar:

> not to invade a country or to subjugate savage peoples, but to form a firm alliance, and to establish ties of friendship and mutual advantage with a new people. Today it is neither with the cross nor the sword that we ought to establish ourselves among these new peoples. It is through respect for their rights and their property, by regard for their customs and traditions that we will win their hearts . . . a conquest of a new type.[16]

Conquest was primarily the issue when the nexus of control and gain was land, not trade. This was the case with settlement colonies, both across the oceans and also those on the land borders of Europe, if such a term can be employed to discuss the borders that European Christian states had with non-Christian powers. The seizure of land might involve profit-sharing, as trade did: it was not necessary to transfer all of the land ownership, and the pre-existing population could maintain a tenurial role. However, the seizure of land generally entailed expropriation and the pauperization of much of the earlier population. The *Domesday Book* (1086) clearly recorded this aspect of the Norman conquest of England.

Expropriation was even more the case if there was a change of usage when land was acquired, in particular a move from nomadic or semi-nomadic pastoralism, or slash-and-burn agriculture, or hunting and gathering practices to settled arable farming. Such a shift was very disruptive, difficult for the earlier population to accommodate, and generally involved violence. At that point, it is possible to present the strength and success of the European (later, Western) military presence as indeed crucial to the purpose of European activity, not that this process was only practised by Western powers.

The Europeans were not the sole peoples and powers to expand. Any explanatory model of the role of force in such expansion instead has to take note of the variety of powers that did expand. Indeed, the Europeans were in many respects distinctive, not because they alone expanded and thus provide a subject for explanation, but because much (although by no means all) of their expansion was trans-oceanic. This was very different to that of other powers such as, in the period 1400–1700, the Mughals, Safavids, Mings and Manchus.

Other powers did have navies. That of the Ottoman Turks was particularly large in the sixteenth century. However, the Turks concentrated their naval operations on galley-borne fleets only capable of relatively short journeys before taking on fresh water and food. Operations such as the attack on Malta in 1565 and the invasion of Cyprus in 1570 involved very large forces. Nothing comparable was deployed in the Indian Ocean. There the Turks took over the Mameluke naval presence in the Red Sea, but unsuccessful attempts to challenge the Portuguese in Indian waters in the 1530s were not sustained, and the Turks did not use their naval presence in the Indian Ocean in order to launch empire-building, let alone colonization. Elsewhere, however, especially in the East Indies and in Hawaii in 1790–1810, non-Western maritime power did serve as the basis for empire-building. Furthermore, the Sultanate of Oman was able to establish a presence on the Swahili coast of East Africa in the late seventeenth century, and to sustain it for two centuries.

Western European exceptionalism can thus be questioned, but there was a distinctive character to maritime states able to project power across the globe, as only the Europeans could do – even if the impact of that power was most effective when directed against other Europeans. The latter were dependent on maritime links and shared the same assumptions about the use of power and the nature of victory. They were therefore especially vulnerable. European power projection was frequently primarily designed to thwart other Europeans. In November 1806, the Foreign Secretary explained the British decision to send troops to Egypt:

> The object of His Majesty's government, in determining upon this measure, is not the conquest of Egypt but merely the capture of Alexandria for the purpose of preventing the French from regaining a footing in that country and of enabling His Majesty's forces there to afford countenance and protection to such of the parties in that country as may be best disposed to maintain at all times a friendly intercourse with Great Britain.[17]

One effect of maritime projection was to bring Europeans into radically different ecologies. As it was, disease was a serious problem for operations on Europe's frontiers, killing large numbers in campaigns against the Turks.[18] The situation further afield, for example in Africa, was more serious, although disease also played a major role in aiding European conquest of the New World. Furthermore, disease hit the armies of other powers. Thus in Taiwan, Chinese forces suppressing rebellion in 1721 and 1786 and Japanese invaders in 1874 and 1895 were both badly affected, especially by malaria.[19]

Aside from the general impact of disease, there were also more specific operational consequences. For example, disease was an important element in deciding how best to combine forces from different areas. The latter was a crucial aspect of the global capability of European maritime empires. In September 1741, when Britain was at war with Spain, Colonel John Stewart wrote from the British colony of Jamaica:

> if ever Britain strikes any considerable stroke in this part of the world the blow must come from the North American colonies not by bringing raw men from thence like those we had last, but by sending officers of experience and good corps to incorporate with and discipline the men to be raised there, these troops as the passage from thence is much shorter might be transported directly to any part of the Spanish West Indies and arrive there with the health and vigour necessary for action, whereas troops sent from home as our own experience has taught us, are by the length of the passage one half disabled with the scurvy and the other half laid up with diseases contracted by confinement and the feeding of salt provisions.[20]

The military articulation of imperial systems was also a matter of the movement of money as much as men.[21]

If models of military success should incorporate non-European forces, then it is also necessary to avoid any treatment of the latter in terms of a hierarchy of capability constructed by similarity to European methods. Such a hierarchy is a classic feature of the primitivization of the military 'other', and such a process, explicit or implicit, lies behind much writing on military history and military power. Thus, to take the example of a first-rate piece of work, *The Cambridge Illustrated History of Warfare*, edited by Geoffrey Parker (Cambridge, 1995), it is only when one turns from cover to title page that the sub-title 'The Triumph of the West' appears. The Preface acknowledges 'the charge of Eurocentrism', but offers:

> three defences. First, it would be impossible to provide adequate coverage in a single volume of the military history of all major

cultures . . . Second, merely to pay lip-service to the military and naval traditions of Africa, Asia, and the Americas, while devoting the lion's share of the attention to the West, would be unpardonable distortion. Finally . . . for good or ill over the past two centuries the western way of war has become dominant all over the world.

All three reasons are understandable, but the net effect is limiting as a coverage of war in the world. It also contributes to a primitivization of non-Western traditions, because clearly they did not lead to a situation of dominance. This underrating is not only a matter of theory, but also of example. Thus, Parker refers to the limited nature of the Turkish military impact in Europe: '. . . Xerxes, Hannibal, Attila, the Arabs and the Turks achieved only short-term success'[22] – a misleading linkage. The Arabs powerfully moulded Spanish culture, while the Turks had an appreciable presence based on conquest for over 500 years: far longer, for example, than the Western presence in Africa or South Asia.

Furthermore, the Turks were never conquered by European powers, and were able to defeat Western armies as late as the 1910s and 1920s: British and Imperial forces at Gallipoli (1915) and at Kut el Amara in Mesopotamia (1916), and the Greeks in Anatolia (1921–2). Such defeats are sometimes explained in terms of the diffusion of Western practices and equipment, but that underrates the strength of non-Western practices and traditions, as well as their capacity to generate change without seeing this as simply a response to Western pressures. Turkey remained neutral in the Second World War, and was too formidable to be intimidated into alliance by either side. In the Cold War, it was regarded as a major addition to the Western bloc.

The Parker volume is far from alone in presenting global military history in a Western frame. *The Reader's Companion to Military History* (New York, 1996), edited by Parker and Robert Cowley, '"privileged" Western matters' on the grounds 'that the Western way of warfare has come to dominate armed conflict all over the globe'.[23] Also in 1996, Cambridge University Press published two volumes of a *Cambridge Illustrated Atlas of Warfare*. The first, *The Middle Ages, 768–1487*, ignored most of the world. Asia was there only for the Crusaders to attack. The second, my own *Renaissance to Revolution 1492–1792*, included coverage of non-Western topics, but very much at my insistence and despite the indifference of the publisher. For example, the initial guidance I was given as to contents stressed the need to give full coverage to the English Civil War (note: not British civil wars of 1637–1652) and the American War of Independence, and there was no interest in including China.

A separate aspect of neglect is more insidious. First, excellent books on war in the West are commissioned and written, for example Michael Howard's *War in European History* (Oxford, 1976), and publishers do not make comparable efforts with other parts of the world, or these areas appear as spheres for the projection of Western power. This is true of significant individual works, for example Brian Bond's *The Pursuit of Victory: From Napoleon to Saddam Hussein* (Oxford, 1996) or Harold Winters' *Battling the Elements: Weather and Terrain in the Conduct of War* (Baltimore, 1998), series such as Longman's 'Modern Wars in Perspective' and Arnold's 'Modern Wars', and also important edited volumes such as John Lynn's *Tools of War: Instruments, Ideas, and Institutions of Warfare, 1445–1871* (Urbana, 1990). Secondly, within the 'West', or the Westernized parts of the world, different military traditions are neglected, and there is a serious process of simplification. For example, accounts of European warfare underplay Eastern Europe. Jeffrey Grey's *Military History of Australia* (Cambridge, 1990; 2nd edn 1999) begins in 1788 with the arrival of British forces and presents the Aborigines in terms of their resistance to British imperialism. Even so, there is only a brief discussion of Aboriginal warfare.[24]

More generally, it is necessary to study non-Western military history – both warfare and military organization – from a non-Western point of view, in order to avoid employing inappropriate methods of analysis. This entails not employing Western military categories, and also seeking to move beyond Western descriptive and analytical vocabularies and methods, and Western constructions of the military character of other cultures and societies.[25]

In short, there is a challenge to historians in East, South and South-East Asia, in Africa, and elsewhere. It is important to ask questions which emerge from local cultures and conditions. Furthermore, it is necessary to search for and use sources relative to military history, but originating from outside the military, in order to understand the role of non-military factors. More specifically, it is necessary not to apply Western distinctions between the military and society.

In assessing warfare, it is important to move away from the suppositions latent or explicit in much published work. It is, for example, necessary to consider conquests by armies that were not the forces of settled societies or advanced states, the last understood in European terms. This is true, for example, of Uzbek pressure on Persia in the sixteenth century and the Afghan campaigns in Persia and north India in the eighteenth. These were all by predominantly cavalry armies, and it is necessary to avoid a perspective predicated on the limitations, even obsolescence, of cavalry, the latter a characteristic of much work on warfare in Europe. Indeed, the value of cavalry, both as a weapons system and as a means to overcome the distances and difficult environments of South and Central Asia, requires attention. Furthermore, cavalry remained important in European operations and battle.[26]

In terms of force–space and force–logistical pressure ratios, cavalry was more effective than infantry across much of Eurasia, and the only comparable weapons system in the pre-industrial world was the wind-driven wooden warship (as opposed to the galley, with its heavy requirement for labour and thus for supplies). The system – the means of delivering, and sustaining force – was more important than the weapon itself, whether bow or gun, for example. This represents a rebuttal of any simple theory of gunpowder triumphalism, and encourages a focus on how gunpowder weapons were adopted and adapted to existing weapon systems. Furthermore, the horse, like the ship, was important as a means for the transport of troops, as well as their supplies. If the ship was the crucial means of transport and supply system for amphibious operations, the horse was scarcely less important to long-range land campaigns.

Just as a more complex approach is offered to the issue of weaponry, so it is also necessary to avoid an overly simplistic account of the geographical and ecological context of warfare. The dichotomy between the West and the 'rest' is limited. It would be helpful to create new, relevant geographical categories that go beyond West and non-West, or land and sea. 'Eurasia', a term employed above, has obvious limitations. For example, there is a difference in the impact of the warhorse and the nomad between the Middle East and Persia. South Asia can be divided in ecological terms between arid and monsoon Asia. This distinction, and others, had a major impact on military logistics and on state-formation.

The need to prepare for different environments remained significant during the long era of Western power projection. The British Military Mission that reported in April 1944 on conflict with the Japanese recommended that:

> Administration and control are so difficult in the jungle-covered undeveloped countries with which we are concerned, that the simplest possible organisation is essential. This organisation must, however, be flexible, because the country varies to a remarkable degree and the nature of the battle will vary in accordance with the terrain. The battlefield varies from dense jungle, where visibility is limited to a few yards, to teak and sal forests where visibility may be a hundred yards. We also find open park land, with large timber and clumps of dense jungle.[27]

Geographical contexts remain important. They are frequently underrated, in part because of the availability today of weapons systems, such as rockets, aeroplanes and modern warships, that are both long-range and, in part, free of geographical constraints. However, the continued role of such factors, even for these systems, was revealed in the 1999 Kosovo conflict, when poor weather affected Allied bombing. On land, the constraints of terrain and, more generally, environment remain very important, although weather is obviously also crucial for both sea and air operations. This is not only the case for Western powers. The Chinese Army, for example, faces very different operating conditions on its lengthy Russian frontier from those on its forested border with Vietnam.

Issues of adoption and adaptation of weaponry further underline the importance of a concentration on organization, and thus cultural issues. In terms of gunpowder weaponry, this leads to an emphasis on drill and discipline,[28] on firing on command, and on operating as units. It is not clear, however, whether this again is an aspect of Eurocentricity. In the case of gunpowder weaponry, the native Americans of New England showed in the seventeenth century that individually aimed fire could be more effective than the volley fire that is stressed in accounts of European warfare.[29] The Maori of New Zealand were to reveal further deficiencies in European methods in the nineteenth century, although the British found it very difficult to recognize the tactical sophistication of Maori warfare. In regard to non-Western peoples generally, this inability seems to be shared by many military historians to this day.

It is also necessary to stress the degree to which different gunpowder weapons had varied implications in terms of organization and socio-cultural contexts, and thus that a more complex account than one based on a simple model of the diffusion of gunpowder weaponry should be offered. For example, whereas earlier muzzle-loading weapons had not been appropriate in East Africa, breech-loaders were far more so because they did not require rigidly disciplined and closely packed bodies of troops, and were thus more compatible socially and culturally with African modes of warfare.[30]

In addition, the drill, discipline and tactics adopted by Europeans to maximize and make best use of their firepower could become institutionalized and persist beyond their point of value, a point that indeed frequently only became apparent in hindsight. This was true, for example, of the continued use of volley fire in European armies at the close of the nineteenth century. In 1898, in the Spanish–American War, the Spaniards fired modern guns – Mausers – in volley on command, while the French in their annual manoeuvres had as their infantry fighting line 'swarms of skirmishers . . . dressing and closed formations are no longer permissible'.[31] The notion that tanks should be used primarily in an infantry support role also had major advocates, not least among German generals, into the early stages of the Second World War. This doctrine, and the related tactics, had obvious consequences for the organization of tank forces.

More generally, if infantry firepower is not seen as the crucial war-winner, then the values associated with its successful use either become less significant or have also to be understood with reference to other weapon systems. For example, it is possible to offer a positive evaluation of the organization, conduct and effectiveness of early-modern African forces. In addition, the role of organization in the widest sense can be questioned: in nineteenth-century Africa, acephalous (stateless) societies proved difficult to conquer, and the same was true of Amazonia. In contrast, societies with more explicit governmental structures provided more easily identifiable targets for attack.

Furthermore, their military was more prone to follow tactics of concentration that offered targets for European firepower. This can be seen in the Sudan campaign of 1898. This is generally seen as a British triumph in which firepower and logistics played crucial roles. That was indeed the case, but in large part this reflected the failure of the Khalifa to make best use of the potential of the Sudan for defence, and his reliance instead on a strategy and tactics that played to the British advantage, not least by permitting force concentration and the use of firepower. A recent study noted:

> By neither threatening Kitchener's line of communications nor attempting to lure him into the hinterland, the Khalifa had greatly simplified the tasks confronting the invading army . . . he was willing to fight a major set-piece battle to defend his capital. . . . The firepower advantage was considerable but its impact was magnified by the strategic and tactical errors of the Khalifa.[32]

The obverse of this was that such an advantage could be lessened by different strategic and tactical choices. In part, there was a learning curve in adjusting to Western warfare, but in some cases the speed and initial impact of the first assault led to a degree of disruption that precluded the effective development of counter-strategies and tactics.

The European conquest of Africa may now appear transitory in world historical terms, although that of Amazonia by European Americans is still continuing, as far as the native population is concerned. This contrast invites consideration of what conquest entails. The total or partial extirpation of a native population, or at least their outnumbering by a settler population granted control, as practised, for example, in the south Russian steppes, Siberia, Australia, Amazonia, Patagonia and the USA, is very different to the creation of a new political authority that has to negotiate with the existing population, even if the terms of that negotiation are established by violence. In the former case, the demographic imbalance was such that the use of force, especially after the initial contact engagements, was a matter not so much of battle, but of what the invaders construed as 'pacification'. Military success rested in part on the ability to exploit differences within the native population, especially between tribes, and on the pace of colonization, with the creation of a supportive settler population and an infrastructure. Settlement was also important where the native population remained numerous, for example with the planting of Scottish Protestants in Ulster and of Russians in the Ukraine in the seventeenth century, and of Han Chinese in Xinjiang in the nineteenth and twentieth centuries. This policy remains important today, as seen, for example, with Chinese moves in Xinjiang and Tibet, Indonesian in Borneo, and Israeli settlements in areas conquered in 1967.

In the case of the conquest of substantial indigenous populations that remain in place, there were again the initial contact engagements, generally on a far larger scale, but thereafter, in many cases, military control and effectiveness rested in large part on the ability to create, or adapt to, structures of authority and consent. Without these, it was difficult to obtain local supplies and also to raise troops, both of which were necessary in order to anchor control. This was true not only of relatively well-known cases, such as the British in India, but also, for example, of the Mughals in north India in the sixteenth century, the Manchu in China and Mongolia, and the Russians in central Asia. This issue was especially important where the demographic balance was heavily against the invader. The examples just given cover the period 1500–1990, and are a reminder that this political dimension to conquest and rule is as long-lasting as any narrower definition of military effectiveness.

Mention of non-European examples ensures that it is possible to qualify any suggestion that the Europeans had a distinctively successful method of control – another variant of the emphasis on organizational superiority and of the notion of Western dominance. Indeed, in the long term, it is unclear whether the Europeans were more successful in this sphere than other peoples. The limited chronological span of their imperial control in Africa, South-West, South and South-East Asia is notable. Over most of these areas, such control was wielded for less than a century, and in some extensive areas, such as northern Burma, northern Nigeria and southern Algeria, even nominally, for sixty years or less.

In most cases, the loss of empire was not a matter of foreign conquest, although the Germans and Italians lost their empires that way in the First and Second World Wars respectively. Instead, it was the failure of incorporating ideologies that is notable. This was true both of colonies and of 'mother countries'. Thus, the desire to reject British control in protectorates or colonies, such as Egypt in the 1920s and India in the 1930s and 1940s, can be matched by the willingness of the domestic political world to accept the loss of empire.[33] More generally, European empires disintegrated after 1945 as much as a result of a lack of political will as of military defeat. This is even clearer if the fall of the Soviet empire from 1989 is included.

It is worth considering the fall of empires for the light it may throw on their rise. The two, of course, are different processes, and in the case of each it is possible to point to a number of aspects and factors, and to their different combination in particular examples. Yet it is also the case that the fall of empires directs attention to the political contexts of military operations, to the need to win and retain local support (as a military as well as a political resource), to the element of morale among the military, and to the difficulty of translating a presence into lasting success. The last is true both of conquests and of subsequently retaining control.

With the passage of time, the European empires may appear more transitory, but that does not necessarily make them the product of chance. Furthermore, it should not necessarily detract attention from the impact of Western military models, a rise of the West as an apparent paradigm of military success. This issue is considered in Chapter Five.

As far as the creation of these empires is concerned, it is useful to look at the ideological factors within Europe that encouraged conquest in the nineteenth century, and to contrast them, for example, with the situation in China, which, at least in the early decades of the century, acted as a satisfied or sated power. It is also important to note that the nature of the nineteenth-century economy was such as to provide Europe with an ability to utilize its natural resources and to profit greatly from the global reach of its trade and capital at a time when the rate of flow of both increased. This facilitated an expansion that can be seen in terms of commercial presence and resource aggrandizement, as well as territorial control.

Thanks to Westernization and the impact of European trade and capital, many of the active military systems of the second half of the nineteenth century throughout the world were not too dissimilar to those of the European powers, as can be seen most obviously in the case of Japan, but also, for example, with Egypt. That does not mean that they were identical, and indeed, the Japanese were to pride themselves on a distinctive martial culture. The development of the Ethiopian Army under Tewodros II, Yohannis IV and Menelik II produced a large force nearly half of whom were armed with modern weapons, but the military culture and political context were different to those in Europe.[34] This was also true of Afghanistan under Abdur Rahman (1880–1901): the army, in part, supported itself by punitive raids. In some cases, indeed, the very process of borrowing may well have encouraged just such a stress on distinctiveness, and it is necessary to employ concepts such as diffusion and borrowing with care.

21

Decolonization greatly increased the number of sovereign military powers from 1946, when the Philippines ceased to be an American colony, and 1947, when India and Pakistan gained independence. Combined with the disintegration of certain states, for example Liberia and Sierra Leone, and the ready availability of weaponry in much of the world, this has ensured that the number of military 'players' is greater today than was the case in the early twentieth century. This increase challenges any attempt to offer a simple model of the military and warfare today. A crude typology can, however, be suggested by emphasizing the contrast between domestic and international demands on the military. The former are becoming more intense, as it becomes harder to elicit and sustain consent in many states, both autocratic and democratic. The withholding of consent, by either the majority or a significant minority, can readily lead to violence, especially if it is linked to a relative lack of tolerance of ethnic and religious diversity.[35] Although the governmental response is presented as policing, it can entail a level of military commitment that deserves the description 'war' – a reminder of the need to take careful note of definitions when discussing conflict.

The intractability of domestic opposition is such that protest is not stilled by the sight of overflying jets and other displays of military power. The technology, training and attitudes developed to deal with international confrontation, frequently presented as the maintenance of peace in the international community, are very different to those that are appropriate for dealing with domestic problems.

The leading military power, the USA, has not had to consider the latter for more than a century, and indeed, the nature of the American Civil War (1861–5) was such that it more closely approximated to an international than a counter-insurgency struggle (see Chapter Six). A central aspect of the American way of war is that the regular armed forces do not police or kill civilians, a situation facilitated by the availability of the National Guard. In the two decades after the Civil War, the National Guard was deployed on 411 occasions to control civil disorder.[36]

Federal troops were sent to Little Rock, Arkansas in 1957 to maintain order during school integration, but in general, discontent and disorder were contained after 1945 without the deployment of such forces. Thus, there are no parallels in recent decades to the marines being ordered to fire into a rioting crowd of xenophobic 'plug-uglies' in Washington in 1857 in order to ensure the proper conduct of an election. Suffering casualties, the crowd fled.[37] Indeed, federal law prohibits the use of active-duty military personnel against civilians without a presidential decree. Despite (or in part possibly because of) the militarization of the police through the provision of training and surplus equipment to Special Weapons and Tactics (SWAT) teams, and despite the use of bellicose language over drugs, the military has not been employed recently for policing in the USA.

This absence of a need to use national troops to deal with domestic opposition, however, is atypical, and ensures that the value of using the USA as the paradigm of the modern military is limited. Over the last century, all other major armies have been used to maintain domestic order, and in the period 1985–2000 this remained the case, with the use of the military in China, Russia and the United Kingdom. Given the widespread practice over the last half-century of eschewing territorial aggrandizement and accepting international boundaries, it is likely that this domestic rationale will remain relatively important, and possibly become more so. This rationale accentuates and focuses another problem with the modern use of the military that also throws light on the role of cultural suppositions. The notion of 'progress' towards a more effective killing and controlling machine is not one with which all, or indeed many, modern societies and commentators are comfortable.

Despite a likely emphasis on domestic policing by the military in many states, intervention in the internal affairs of other states in support of domestic opponents, for example Pakistani support of Kashmiri secessionists, or Turkish, Iraqi and Iranian participation in the Kurdish question, may be such that the inability to secure domestic compromises leads to international conflict, as happened in a brief and limited fashion between India and Pakistan in 1999.

It is unclear in these and other cases how far policies are and will be set with reference to military advice and an informed awareness of military capability. Political leaders are apt to promote and listen to those who provide the advice that they seek.[38] Initial commitments are frequently made in a secret, sometimes illicit, fashion, and then troops are committed in order to support the commitment. The most worrying aspect is the widespread ignorance in both governments and public opinion about the nature of war and the problems of securing and maintaining victory. If only for this reason, military history has a very valuable role to play.

This was driven home in 1999, when George Robertson, the British Secretary of State for Defence, publicly scorned military historians who had warned about the difficulty of winning the Kosovo conflict by air power alone, and also about the contrast between output (bomb and missile damage) and outcome. When the subsequent Serbian withdrawal revealed that NATO estimates of the damage inflicted by air attack, for example to Serb tanks, had been considerably exaggerated, Robertson did not apologize. Indeed, he was soon after appointed Secretary General of NATO and gained a peerage. In 1922, the General Staff of the British Forces in Iraq had observed, in a military report on part of Mesopotamia: 'Aeroplanes by themselves are unable to compel the surrender or defeat of hostile tribes.'[39]

NOTES

1. J. Lynn, 'The Embattled Future of Academic Military History', *JMH*, 61 (1997), pp. 777–89. This is essentially about the USA. See also V. Hanson, 'The Dilemma of the Contemporary Military Historian', in E. Fox-Genovese and E. Lasch-Quinn, *Reconstructing History* (1999), pp. 189–201, esp. 190, 196–7.
2. P.B. Evans et al. (eds), *Bringing the State Back In* (Cambridge, 1985).
3. B.M. Downing, *The Military Revolution and Political Change: Origins of Democracy and Autocracy in Early Modern Europe* (Princeton, NJ, 1993); T. Ertman, *Birth of the Leviathan. Building States and Regimes in Medieval and Early Modern Europe* (Cambridge, 1997); S. Rosen, *Societies and Military Power: India and Her Armies* (Ithaca, NY, 1997).
4. V. Hanson, *The Western Way of War: Infantry Battle in Classical Greece* (New York, 1989); T. Farrell, 'Culture and Military Power', *Review of International Studies*, 24 (1998), pp. 405–14.

5. M. Roberts, *The Military Revolution, 1560–1660* (Belfast, 1956).
6. For the naval situation, and an awareness of theory, see J. Glete, *Navies and Nations. Warships, Navies and State Building in Europe and America, 1500–1860* (Stockholm, 1993).
7. G. Hanlon, *The Twilight of a Military Tradition: Italian Aristocrats and European Conflicts, 1560–1800* (1998).
8. L. Campbell, 'Recent Research on Andean Peasant Revolts, 1750–1820', *Latin American Research Review*, 14 (1979), pp. 3–50; O. Cornblit, *Power and Violence in a Colonial City: Oruro from the Mining Renaissance to the Rebellion of Tupac Amaru, 1740–1782* (Cambridge, 1995).
9. F.C. Lane, *Profits from Power: Readings in Protection Rent and Violence-Controlled Enterprise* (Albany, NY, 1979).
10. T.J. Barfield, *The Perilous Frontier: Nomadic Empires and China, 221 BC to AD 1757* (Oxford, 1989); J. Gommans, 'The Silent Frontier of South Asia, c. AD 1100–1800', *Journal of World History*, 9 (1998), pp. 1–23.
11. G. Cornwallis-West, *The Life and Letters of Admiral Cornwallis* (1927), pp. 50–1.
12. R. Jervis, *Perception and Misperception in International Politics* (Princeton, NJ, 1976); A.I. Johnston, *Cultural Realism: Strategic Culture and Grand Strategy in Chinese History* (Princeton, NJ, 1995).
13. See, in particular, D. Ralston, *Importing the European Army: The Introduction of European Military Techniques and Institutions into the Extra-European World, 1660–1914* (Chicago, IL, 1990).
14. Walter Titley to Edward Weston, 23 May 1761, Farmington, CT, Lewis Walpole Library, Weston papers vol. 21. More generally, see, A. Wendt, 'Anarchy is What States Make of It: The Social Construction of Power Politics', *International Organization* 46 (1992), pp. 391–425.
15. A. Reid, *Europe and Southeast Asia: The Military Balance* (Townsville, Queensland, 1982), and *Southeast Asia in the Age of Commerce 1450–1680, Volume II: Expansion and Crisis* (New Haven, CT, 1993), pp. 87–90; J.A. Brandão, 'Your Fyre Shall Burn No More': Iroquois Policy toward New France and its Native Allies to 1701* (Lincoln, NB, 1997).
16. *Archives parlementaires de 1787 à 1860: Recueil complet des débats législatifs et politiques des chambres françaises* (127 vols, Paris, 1879–1913), pp. 37, 152.
17. Howick to General Fox, 21 November 1806, BL Add. vol. 37050 fol. 46.
18. Charles of Brunswick Bevern, writing from Belgrade, to Frederick William I of Prussia, 24 September 1738, Wolfenbüttel, Staatsarchiv, 1 Alt 22.549 fol. 230.
19. P.R. Katz, 'Germs of Disaster: The Impact of Epidemics on Japanese Military Campaigns in Taiwan, 1874 and 1895', *Annales de Démographie Historique*, 1 (1996), pp. 196–220. For attempts to work disease into the wider picture, see, for example, A.W. Crosby, *The Columbian Exchange: Biological and Cultural*

Consequences of 1492 (Westport, CT, 1969) and *Ecological Imperialism. The Biological Expansion of Europe, 900–1900* (Cambridge, 1986).

20. Stewart to Earl of Stair, 10 September 1741, New Haven, CT, Beinecke Library, Osborn Shelves, Stair Letters, no. 70.
21. C. Marichal and M.S. Mantecón, 'Silver and Situados: New Spain and the Financing of the Spanish Empire in the Caribbean in the Eighteenth Century', *Hispanic American Historical Review*, 74 (1994), pp. 587, 589, 597–8, 606–10.
22. G. Parker (ed.), *The Cambridge Illustrated History of Warfare* (Cambridge, 1995), pp. vii, 9.
23. R. Cowley and Parker (eds), *The Reader's Companion to Military History* (New York, 1996), p. xiii.
24. J. Grey, *A Military History of Australia* (2nd edn, Cambridge, 1999), pp. 29–31. See also the focus on the American (European-American) experience in R. Doughty et al., *Warfare in the Western World* (2 vols., Lexington, MA, 1996).
25. For the last, see J. Lamphear, 'Brothers in Arms: Military Aspects of East African Age-Class Systems in Historical Perspective', in E. Kurimoto and S. Simonse (eds), *Conflict, Age and Power in North East Africa: Age Systems in Transition* (Oxford, 1998), pp. 79–97 and D. Harrison, *Social Militarisation and the Power of History: A Study of Scholarly Perspectives* (Lund, 1999).
26. For cavalry in Africa, see R. Law, *The Horse in West African History: The Role of the Horse in the Societies of Pre-Colonial West Africa* (1980). On Europe, see, for example, G. Phillips, *The Anglo-Scots Wars 1513–1550* (Woodbridge, 1999), pp. 27–30.
27. PRO WO 33/1819, p. 40.
28. See H. Kleinschmidt, 'Using the Gun: Manual Drill and the Proliferation of Portable Firearms', *JMH*, 63 (1999), pp. 601–29.
29. Patrick Malone, *The Skulking Way of War: Technology and Tactics Among the New England Indians* (Lanham, MD, 1991).
30. J. Belich, *The New Zealand Wars and the Victorian Interpretation of Racial Conflict* (Auckland, 1986); J. Lamphear, 'The Evolution of Ateker "New Model" Armies: Jie and Turkana', in K. Fukui and J. Markakis (eds), *Ethnicity and Conflict in the Horn of Africa* (1994), p. 89.
31. PRO WO 32/2819, p. 12.
32. E.M. Spiers, *Wars of Intervention: A Case-study – The Reconquest of the Sudan 1896–99* (Camberley, 1998), pp. 46–7. For another case study, the success of the British expedition of 1918 against the Turkana of Kenya, see J. Lamphear, *The Scattering of Time: Turkana Responses to Colonial War* (Oxford, 1992).
33. J. Darwin, 'The Fear of Falling: British Politics and Imperial Decline since 1900', *Transactions of the Royal Historical Society*, 5th series, 36 (1986), pp. 27–43.

26

34. H.G. Marcus, *The Life and Times of Menelik II: Ethiopia, 1844–1914* (Oxford, 1975); J. Dunn, '"For God, Emperor, and Country!" The Evolution of Ethiopia's Nineteenth-century Army', *War in History*, 1 (1994), pp. 278–99.

35. K.J. Holsti, *The State, War, and the State of War* (Cambridge, 1996) and *Political Sources of Humanitarian Emergencies* (Helsinki, 1997); G. Sørensen, 'An analysis of Contemporary Statehood: Consequences for Conflict and Cooperation', *Review of International Studies*, 23 (1997), pp. 253–69.

36. J. Cooper, *The Rise of the National Guard: The Evolution of the American Militia, 1865–1920* (Lincoln, NB, 1997).

37. J.G. Dawson, 'With Fidelity and Effectiveness: Archibald Henderson's Lasting Legacy to the U.S. Marine Corps', *JMH*, 62 (1998), pp. 750–2.

38. For a powerful critique from a serving officer, see H.R. McMaster, *Dereliction of Duty: Lyndon Johnson, Robert McNamara, the Joint Chiefs of Staff, and the Lies that Led to Vietnam* (New York, 1997), for example pp. 322–5. See also R. Buzzanco, *Masters of War: Military Dissent and Politics in the Vietnam Era* (Cambridge, 1997).

39. Military Report on Mesopotamia (Iraq) Area 1 (Northern Jazirah), 1922, PRO WO 33/2758, p. 39. The extensive literature on this topic includes D.J. Mrozek, *Air Power and the Ground War in Vietnam: Ideas and Actions* (Maxwell Air Force Base, AL, 1988), M. Clodfelter, *The Limits of Airpower: The American Bombing of North Vietnam* (New York, 1989), D.M. Drew, 'Air Theory, Air Force, and Low Intensity Conflict: A Short Journey to Confusion', in P.S. Meilinger (ed.), *The Paths of Heaven: The Evolution of Airpower Theory* (Maxwell Air Force Base, AL, 1997), G.K. Williams, *Biplanes and Bombsights: British Bombing in World War I* (Maxwell AFB, 1999), C.C. Crane, *American Airpower Strategy in Korea, 1950–1953* (Lawrence, KS, 2000) and Meilinger, 'The Historiography of Airpower: Theory and Doctrine', *JMH*, 64 (2000), pp. 467–502.

TWO

The Development of Military Organizations to 1850

The cavalry regiments . . . attract officers from families which in the past have preserved the feudal conception that the holding of estates carries with it a liability for defence of the kingdom.

Report of the [British] Committee on the Mechanized Cavalry and Royal Tank Corps, 1938.[1]

Much of the scholarly work on military history has adopted an explanatory model of change that centres on the impact of new military technology. This can be readily related to operational accounts of war, and also offers a method that can be used both to cover the entire world and to explain shifts in the relationship between different parts of the world. The principal alternative approach to military history that is currently fashionable focuses on the so-called 'new military history' (now no longer new). The emphasis here is on social contexts, especially the position, experience and relationships of the rank and file.[2] The 'new military history' is important to any understanding of the character of military forces. Furthermore, probing the dynamics of units under operational circumstances, whether in conflict or not, is crucial to an understanding of their capability and effectiveness, not least to the question of why forces and units equipped with similar weapons can enjoy very different rates of success. The 'new military history', however, is less helpful from the perspective of providing a long-term explanatory model of military change.

This chapter and the next adopt another approach, that of military organization. This can be regarded as important in itself, and also as a valuable perspective from which to consider issues of military success and change, and to probe the causes of war.

28

'Organization' can be understood in a double sense: first, the explicit organization of the military – unit and command structures – and second, organization as an aspect of, and intersection and interaction with, wider social patterns and practices, leading to the social systematization of organized force. Attitudes towards hierarchy, obedience and discipline, the readiness to serve and the willingness to pay taxes to support war and the military, not least on the part of the privileged, also have to be understood in cultural as much as functional terms. These two senses of 'organization' are not completely separable; indeed, were they to be so, military units would not be effective for long. Nevertheless, they permit an understanding of the subject that reflects the multiple character of military organization, while focusing on the specific nature and unique function of armed forces.

Another important aspect of organization is provided by the systematization of knowledge, such that it is possible better to understand, and thus seek to control, the military, its activities and its interaction with the wider world. Such a process was true, for example, of statistics and mapping in the early-modern Western world. Resources could be located. In 1764, William Harrison's chronometer lost only five seconds when HMS *Tartar* crossed from Portsmouth to Barbados. Knowledge was also institutionalized, and the necessary organizations were developed accordingly.[3]

The origins of modern military organization are not chronologically precise. The first crucial dimension of the dynamic of social change in and through war occurred when the military ceased to be coterminous with society, more specifically with adult male society. This ensured that the organization of this society for war was replaced by a situation in which a segment of society was thus organized, while the rest of society was organized to provide for, or at least provide, this segment. If modern military organization is understood primarily as a professional force of trained regulars under the control of sovereign powers, and with such powers enjoying a monopoly of such forces, then the chronology and explanation of such development varies greatly. The situation was very different in, say, Siberia and Mediterranean Europe.

29

Furthermore, any understanding of military organization has to be wary of a state-centred, let alone Eurocentric, perspective, whether in definition, causality or chronology. Many military organizations have not been under state control, or were under only limited control,[4] as is clear from any discussion of conflict today. It is necessary to be cautious about assuming a teleological, let alone triumphalist, account of state control of the military. Indeed, it is questionable how far such a monopolization should be seen as an aspect of modernity.

In addition, it is appropriate to note the problematic character of modernity as a concept, whether descriptive or prescriptive – an issue discussed, for the early-modern period, in Chapter Four. Aside from the role of modernization as a polemical device in political debate, there is, in analytical terms, a difficulty in determining how best to define and dissect the concept. The triumphalism that saw it in terms of the rise of mass participatory democracy, secular, or at least tolerant, cultures, nation states, and an international order based on restraint has been eroded by a series of critiques, from both within and outside the West.

This has a direct bearing on the understanding of military organization. Thus, for example, conscript armies could be seen in progressive terms in the nineteenth century, as an adjunct of the extension of the male franchise. Both symbolized a new identification of state and people in countries such as France, Germany and Italy, although not in Britain or the USA. In the twentieth century, conscription was also important in the ideology of Communist states. In addition, after 1945, as a new politics was created in what had been fascist societies, conscription was seen, for example in Germany and Italy, as a way to limit the allegedly authoritarian and conservative tendencies of professional armies, particularly their officer corps.

In the more individualistic Western cultures of the 1960s onwards, however, conscription as a form, rationale and ideology for the organization of the military resources of society seemed unwelcome. Military service was presented less in terms of positive images, such as incorporating ideology and social mobility, and more as an unwelcome chore and a form of social control. Conversely, in Latin America, conscription had a more positive image (though far from universal), in part because military service was seen as a means for useful training, and for economic opportunity and social improvement, both for individuals and for society.

Such points underline the contingent nature of judgements, and the danger in adopting a teleological account of change. They also serve as a warning against adopting what may appear to be a self-evident proposition, namely that the purpose of the military is to win wars, and that military organization is designed to improve the chances of doing so. Such a proposition is flawed, and even if the prime emphasis is on war-winning, it is necessary to explain the processes by which such an emphasis affects the operation and development of such organizations.

To take the former point, military organizations serve a number of functions, some of which are publicly defined and endorsed, while others are covert, implied or implicit. National security is an aspect of the former, while employing people and providing possible support for policing agencies are examples of the latter. Far from there being any uniformity in situation or development as far as these various functions are concerned, they vary both in an objective fashion, by state, period and branch of the military, and in a more 'subjective' sense, with reference to the views of leaders, groups and commentators within and outside the military. Thus, French army operations against large scale smuggling gangs in the frontier region of Dauphiné in 1732[5] could be classed as policing or national security, or both. The same is true of modern operations against drug producers and smugglers, for example by American and British military units.

This instrumentality of the military is a matter not only of defining which purpose it is necessary to be fit for, but also extends to the very character of the military organization in a particular society.[6] In other words, the purpose of the organization may not be that of achieving a specific military outcome, or the potential for such an outcome. Rather, the prime objective of the creators of the organization may be to produce a body or system that fulfils and represents certain domestic socio-political goals. This also needs to be considered when assessing the likely future of the military.

Politicians, for example, may be more concerned to ensure 'democracy' in the armed forces, or republican values, or revolutionary zeal, or commanders who will or will not automatically obey the government than they are to consider the war-making potential and planning of the armed forces. Indeed, the latter may be left to the professional military, provided the control culture and value system that are sought are in place.

In 1782, the 3rd Duke of Richmond, Master General of the British Ordnance, wrote to Major General Charles Grey about how to repel a possible French attack on Plymouth. He noted:

> I am sure you can have no idea of the many real difficulties that exist and prevent one's doing business with that dispatch that could be wished. I have many delays to surmount in my own office, but depending also upon others, upon the Commander in Chief who has his hands completely full and then upon a numerous Cabinet which is not the more expeditious for consisting of eleven persons who have each their own business to attend to.[7]

Richmond therein captured the central feature of British military organization arising from the state's politics: the military answered to civilian control, and was expected to do so in war as well as in peace. Discussion of the organization and role of the military in the early American republic reflected political views on republicanism and federalism. Similarly, in 1924 the left-wing government that gained power in France was more concerned about the ideological reasons for shortening conscripts' terms of service than about preparing for war with Germany.

In the modern West, operational military control and political direction are in theory largely disaggregated, although the distinction is hard to maintain, as was discovered in peace-keeping work, for example in Bosnia and Kosovo; and a combination of the speed of modern communications and the pressures of public accountability in democratising societies facilitates and encourages political interference in operational issues. This is understandable, as the course and consequences of operations can have major political consequences.

A separation of military from political responsibilities and power is not generally the situation outside the West. In some countries, such as Iraq, political direction extends to operational control, and in others, such as Burma, senior military figures run the state, or, as in Pakistan and Indonesia, are autonomous, almost a state within a state. Historically, an aggregation of political and military aspects has been more common than their separation. This helps answer the question of how far concerns about war-winning affect the development of military organizations, because these concerns have generally been seen through the prism of political assumptions, and/or have been subordinated to these assumptions.

These theoretical points serve as a prelude to a brief consideration of the chronological development of military organization. At the outset, it is best to consider a sociology of different military systems. The evolution of specialized forces – of trained regulars – under the control of 'states' occurred initially against the background of a world in which there was a general lack of such specialization, and, even more, of such forces, because many specialized units were mercenaries and not, therefore, under state control. In addition, the absence of powerful sovereign authority across much of the world was such that, for 0–1500 CE, it is commonly more appropriate to think of tribal and feudal military organization, rather than a state-centric system. Furthermore, diversity was evidence of the vitality of different traditions, rather than an anachronistic and doomed resistance to the diffusion of a progressive model.

Diversity owed something to environmentalism: the interaction of military capability and activity with environmental constraints and opportunities. Thus, there were different options depending on whether horses were present or absent. This changed to a certain extent with the diffusion of horses, but they could not be used in some areas, for example the tsetse-fly belt of Africa or the mountainous terrain of Norway; whereas in others, such as Hungary, cavalry could operate easily.

Any global chronology of military organization faces serious qualification, because in areas of developed state power, such as China and imperial Rome, professional, state-controlled forces long preceded the middle of the current millennium, whereas in other parts of the world they were created, or imposed, far more recently. Similarities between the military arrangements of the Roman and early Chinese empires included sophisticated logistical systems, and the use of military colonies and of non-native/imperial troops. The relatively low productivity of pre-nineteenth-century agrarian economies was not incompatible with large forces at the disposal of such states, while the constraints that primitive control and command technology and practices placed on centralization did not prevent a considerable measure of organizational alignment over large areas, as also, for example, in the case of the Inca empire in South America in the fifteenth century. The size of some of these forces could be considerable. In China in the Warring States period (403–221 BCE), improved weapons and the use of mass infantry formations led to some of the largest military engagements yet recorded. The Byzantine (Eastern Roman) empire had an army of about a third of a million in the mid-sixth century, and about a quarter of a million in 1025, approximately 1.5–2 per cent of the empire's manpower. There was also a navy. This was a heavy burden that took an average of 69 per cent of the government's budget. However, the size of the army was far smaller after the crisis of 1204, when the Fourth Crusade captured Byzantium.[8]

The situation in such empires was very different to tribal warfare societies, where force was an expression of collective social power, rather than the authority of the state. States could be resilient and could survive conquest by tribal warfare societies. For example, conquests of China were not followed by its collapse. Instead, the Mongols in the thirteenth and the Manchus in the seventeenth century made a transition in which they acculturated to the society they had conquered[9] that was more complete than that of the sixteenth-century Mughal invaders of northern India.

In both Han China and imperial Rome, although central government sought to monopolize force, there was an ambiguous relationship between, on the one hand, recruitment and the power of the socially privileged, and on the other, the governmental desire for a demilitarized interior. China was repeatedly to suffer from the problems created by the rise of internal disorder and the failure of central government to suppress it.[10] Furthermore, in both China and Rome, the governmental desire to monopolize force still faced the organizational problem of controlling frontier units.

The prestige of imperial states, especially China and Rome, was such that their military models considerably influenced other powers, especially the successor states to the western Roman empire. Thus, for example, Roman military doctrine was used skilfully by the Frankish army of Burgundy in the 580s. The prestige of Rome revived in the early-modern period, and its military vocabulary was in part employed anew, while Roman military texts, such as Vegetius' fourth-century *De Re Militari*, were printed and referred to.[11]

However, part – frequently much – of the success of both imperial states rested on their ability to co-opt assistance from neighbouring 'barbarians'. They brought valuable skills with horse and bow. This co-option became more marked in the Roman empire as demilitarization of the interior became more pronounced. The role of 'barbarians' ensures that any account of Roman or Chinese imperial military organization that offers a systematic description of the core regulars is only partial. Indeed, both imperial powers deployed armies that were, in effect, coalition forces.

This remained the case with most major armies until the age of mass conscription in the nineteenth century, and even then, was true of their trans-oceanic military presence. Such co-option could be structured essentially in two different ways. It was possible to equip, train and organize ancillary units like the core regulars, or to leave them to fight in a 'native' fashion. Both methods were followed by imperial powers, such as the Romans, or the British in eighteenth-century India.

The net effect was a composite army, and such an organization has been more common than is generally allowed. More generally, the composite character of military forces essentially arises both

35

from different tasks and from the use of different arms in a co-ordinated fashion to achieve the same goal: operational success. Thus, cavalry and infantry, light and heavy cavalry, pikemen and musketeers, frigates and ships of the line, tanks and helicopter gunships combined to co-operate, creating problems of command and control that affected organizational structures, and tactical issues that dominated military doctrine. Skill in command frequently related to obtaining successful co-operation between different arms. For long, the central problem was that of obtaining a successful co-ordination of infantry and cavalry.

The composite nature of large forces referred to above is an aspect of this co-ordination of different arms because, for example, in imperial Rome the native ancillary units commonly provided light cavalry and light infantry to assist the heavy infantry of the core Roman units. Similarly, Mongol and Manchu cavalry co-operated with Chinese infantry. The Ottoman Turks were provided with light cavalry by their Crimean Tatar allies, their Russian enemies by the Cossacks and, in the nineteenth century, by Kazakhs too.

Such co-operation rested not so much on bureaucratic organization as on a careful politics of mutual advantage[12] and an ability to create a sense of identification. Chinese relations with nomadic and semi-nomadic peoples of the steppe always combined military force with a variety of diplomatic procedures – one of the best-known being *jimi*, or 'loose rein', which involved dividing and ruling. The politics of mutual advantage affected patterns of command, frequently by ensuring that 'native' forces operated as a parallel force with no command integration other than at the most senior level. The frequent combination of 'native' cavalry and 'core' infantry suggests that, in part, such military organization bridged divides that were at once environmental, sociological and political.[13] This bridging complemented the symbiotic combination of pastoralism and settled agriculture that was so important to the economies of the pre-industrial world.

Any form of social determinism in military organization ascribing, for example, particular systems to nomadic or industrial societies is limited, even suspect, not least because professionalization in warfare reflected, in large part, cultural constructions of war and military service. These are not simply socially determined. Indeed, notions and practices of service, duty and discipline had to combine in order to make military specialization and hierarchy possible and effective.

In tribal societies, this process was challenged by customs and ideas that were more contractual and, in some respects, egalitarian. There was, in some respects, a conditionality in, for example, the steppe tribes of Central Asia or the native peoples of Native America that was very different to patterns of military control in societies, such as Ming China, that had a more authoritarian political ideology and practice. Native American use of weapons and tactics allowed for much more autonomy on the part of individual warriors than was the case with Chinese soldiers,[14] although it is important not to underrate the role of organized tactics and discipline among Native Americans, nor of leadership by officers.[15] Furthermore, the democratic system and ethos of many (but by no means all) tribal groups helped to maintain their powers of resistance, as the British discovered on the North West Frontier of India with tribes such as the Afridis and Wazirs.

The relationship between imperial societies and military organization was a matter not only of the expansion of the control of such societies over differently organized peoples, but also of changes in the structure of imperial societies. The latter were important not simply in the recruitment and discipline of rankers, but also of the 'officers'. The latter posed serious problems in many societies, because the 'officers' were the socially prominent, and thus those most likely to pose issues of political control. This could influence the organization and use of the military. At present, the basis for a comparative study of this subject is lacking. The nature of military control in many societies is more obscure than that of the weaponry employed.

The most valuable and sustained attempt to develop a taxonomy and analysis of the development of military organization came in an article of 1996 by John Lynn. In an impressive study, Lynn considered 'The Evolution of Army Style in the Modern West,

800–2000', concluding that there were seven types: feudal; military-stipendiary; aggregate-contract; state-commission; popular-conscript; mass-reserve, and volunteer-technical. Lynn skilfully related these types to civilian-military relations, and to the more general issue of state-building. Although he offered a bold sweep, Lynn focused on armies, and was also Eurocentric:

> This study centres on armies, as opposed to armed forces, because armies reflect a society more completely and have the potential to influence it more profoundly. Consider that navies and air forces make poor tools for internal control, *coup d'état*, or revolution, whereas armies are expert at all three. . . . These pages focus on Europe and its projections, most notably the United States, because the characteristics and mechanisms to be discussed here assume a cultural and geographical pattern unique to the West . . . there *were* distinctly Western ways of constructing an army . . . the emphasis on discipline, drill, and ability to suffer losses without losing cohesion would also appear to be a Western trait. . . . Because the Western military style rested upon certain political, social, economic, and cultural foundations – what will be called 'state infrastructure' here – Western states produced particular kinds of military institutions. As other, non-Western, states tried to adopt Western army styles, they found it necessary to import certain aspects of that infrastructure as well.[16]

Yet it is also important and profitable to consider differently organized non-European forces. For example, both Ottoman and Western sources tend to see the Ottoman (Turkish) soldiery of the sixteenth and seventeenth centuries as well equipped and well fed. They were also widely successful. More generally, the view that non-European powers should feature only if they replicate European developments is an unhelpful way to cover the period up to 1750, and indeed 1800, by which time European dominance was still limited in Africa, and in East, South-East and South-West Asia. The Europeans also encountered serious problems in South America.[17]

38

An emphasis on the Western model also omits the extent to which non-European societies have followed different military trajectories in the nineteenth and twentieth centuries, especially if due heed is paid to social and political contexts. In the nineteenth century, for example, China was affected by the same process of Westernization as Japan, but the political context in China was very different, and the response to Western examples more uneven and hostile.

Thus, models devised to discuss and explain the role of the military in modern Europe are unhelpful when it comes to considering China or Iraq, Indonesia or Pakistan. And yet the military in the latter played, and play, a greater role in politics, in government, in the maintenance of control, and in projects of modernity and modernization than their counterparts in western Europe or the USA.

The neglect of the wider world is also serious because it affects our understanding of European military history, specifically by leading to a limited understanding of the problems facing past European military systems. Lynn, like earlier workers in the field, for example Michael Howard,[18] essentially offered a paradigmatic model, and one that centres on conflict *between* European states, particularly in the France-Germany region. This is an account of the pressures of symmetrical warfare, a point emphasized by Lynn's opening example, drawn from the battle of Rossbach in 1757 between Prussia and France. Like the battles of Rocroi (1643) and Valmy (1792), between France and Spain and Revolutionary France and Prussia respectively, Rossbach has an almost totemic significance in accounts of military development, but this is a significance that is exaggerated.

A concentration on paradigmatic conflict between European states is unhelpful as an account both of Europe's military position in the wider world and, indeed, of European warfare itself. A paradigmatic approach *may* be appropriate for certain periods, but is generally unhelpful. To take an example that brings these points together, Lynn supposes that Prussia's success in the German Wars of Unification (1864–71) made the German army a paradigm for other European forces.[19] There is indeed a considerable degree of truth in this, as seen most obviously in the development elsewhere of general staffs consciously modelled on the Prussian example. The fashion for German-style military uniforms was a testimony to Prussian prestige.

However, the German army was also eccentric to the general European military development. Between 1840 and 1885, Britain, France and Russia all had to confront the operational and organizational problems posed by bitter resistance to the advancing frontiers of European control in, for example, India and South Africa, Algeria and Indo-China,[20] and the Caucasus, Central Asia and the Balkans,[21] respectively. Their armies had broader-spectrum missions than that of Germany, and required a correspondingly flexible – or at least differentiated – organization and ethos. These challenges were, in part, addressed by developing particular military practices and institutions that relied on the use of large numbers of native troops. Indeed, more generally, any taxonomy and analysis of military style has to be able to comprehend asymmetrical as well as symmetrical warfare.

This is true of Europe's military history as a whole, as it is also, for example, of that of 'India', although the latter is very much an abstraction in so far as military history is considered. Over the last two millenniums, and indeed earlier, there have been repeated attacks on European states by peoples from 'outside' Europe, the latter understood in terms of what was to be the Romano-Christian nexus. The leading military challenges in the eighth and ninth centuries were thus Arabs, Magyars and Vikings, while, in the thirteenth century, the Mongols were a major challenge in Eastern Europe, to be succeeded, from the fourteenth to the eighteenth, by the Ottoman Turks. It is, of course, difficult to decide how best to include in this analysis peoples such as the Bulgars in the ninth and tenth centuries or the pagan Lithuanians in the High Middle Ages, because the concept, and reality, of Europe were plastic, not rigid. However, an emphasis on asymmetry between European and non-European warfare is generally appropriate for much of the last millennium. Any paradigm, organizational or otherwise, based solely on inter-European military competition is thus questionable.

This analysis can be taken further by suggesting that such conflicts – on the geographical 'periphery' in Lynn's terms – had a greater impact on the 'core' than a model that privileges developments within the latter might suggest. This is true not only of the European military system, in so far as such an abstraction can be employed, but also of those elsewhere, for example in the medieval Islamic world. In short, methods and organizations devised to deal with threats and challenges from outside might well have affected not only 'peripheral' states – a term that for Europe in the seventeenth century included Venice, Austria, Poland and Russia – but also those further removed from the direct confrontation with 'non-European' states, of which the Turkish empire was the foremost. In the War of Ferrara of 1482–4, Venice used Albanian and Greek *stradiots* (light cavalry troops), who had initially been employed to fight the Turks in the Morea and Friuli. Venice's opponents used squadrons of Turkish cavalry captured and re-employed after the fall of Friuli. Austria and Poland acted as intermediaries between modes of conflict designed for warfare on the periphery and those in the core in the seventeenth century, not least in maintaining a role for light cavalry.[22] The impact of such intermediary roles is difficult to assess, because the processes by which armed forces evolve or reform themselves are far from certain. In particular, the respective effects of external challenge and domestic processes are unclear.

In part, this issue of military core-periphery within Europe, in particular of the leading role ascribed to the core, is an aspect of the primitivization of the East in Western analysis, a primitivization that also greatly affected, and affects, perceptions of Eastern Europe in a misleading fashion.[23] More generally, it leads to an analysis in which Westernization is crucial to the development of non-Western societies, and aspects of their military history that point in different directions are ignored or treated as anachronistic.[24] Thus, for example, it has been argued that the continued value and importance of cavalry in eighteenth-century India is neglected.[25] More generally, cavalry is slighted as a military tool, let alone an aspect of the wider politics of army organization. In the latter case, cavalry, not least as a consequence of the role it offered for the landed elite, was important to social cohesion within armies, and between them and the rest of society.

Aside from core–periphery, with the latter understood, in part, as a zone of reaction to, and possibly transition with, non-Western systems, there were also important variations within Europe. Just as medieval Celtic warfare had important differences to its feudal counterpart, so, in the early modern period, Scots and Irish forces using shock tactics occasionally defeated English armies operating in a more conventional European fashion.[26] The same might have been true of the Swiss had they continued to play a major role in power politics after the 1510s.

As has been done for India, the notion of internal military frontiers within Europe might profitably be deployed as an approach to issues of capability, operational activity and organization; and it would also be useful to discuss when and how these frontiers ceased. The issue of variety within Europe can also be related not only to the question of whether it is appropriate to think of distinctive organizational characteristics in particular periods,[27] but also to whether the issue of standing versus contract armies in the medieval period can be better seen both in terms of a continuum and as prefiguring the more recent conscript/professional duality.

Any emphasis on naval developments adds another geographical challenge to the core concept, in this case, from the Classical period, a Mediterranean core, and from the fifteenth century, a Western European core. An inclusion of the naval dimension also poses the problem of determining whether there was a distinctive typology and chronology of naval organizational development.

This was suggested by Nicholas Rodger in his important study of English naval development. Rodger argued that land-based military systems require and encourage an authoritarianism, in organization and political context, that is inappropriate and, indeed, ineffective in the case of naval power. Instead, he suggested that, in the case of the United Provinces (Dutch, modern Netherlands) and England, later Britain, it was the creation of consensual practices and institutions in the early-modern period that was crucial in producing the readiness to sustain naval power. He regarded a relatively open and participatory society as a precondition for naval efficiency, and argued that the social politics and organizational requirements of armies were different, with numbers being the prime requirement and rationale for the latter. Politically, armies can more easily hold down societies, by direct force or its threat, than navies.[28] In fact, although this was not Rodger's subject, size was not the sole measure of effectiveness: some armies were markedly better than others.

Rodger's approach to naval history, like that of many other scholars concerned with naval power, for example Peter Padfield in his *Maritime Supremacy and the Opening of the Western Mind: Naval Campaigns that Shaped the Modern World 1588–1782* (1999), excluded all non-European naval powers. Yet, in the period 1400–1600 alone, China, Korea, Japan and Turkey all deployed major naval forces, while lesser forces were controlled by, for example, Aceh in Sumatra. Even within Europe, although operational and organizational efficiency are difficult to measure, the French navy under Louis XIV (1643–1715), the epitome of absolutism, was equal, or close to equal, in quality to the Dutch or English navies, although outnumbered by the two together; and the Danish navy did not deteriorate after 1660, when Denmark became more 'absolutist'. Autocratic Russia developed a large navy in the eighteenth century, and it remained important thereafter.[29]

Conversely, the open and highly participatory Dutch society created a large and efficient army that was successful against Spain in the seventeenth century. The theory that representative assemblies made it easier to mobilize resources can be sustained in the case of the Dutch and Sweden, but runs into difficulties with Prussia, which didn't have one, and Poland, which did. It is also necessary to consider the counterfactuals that might be advanced in the case of English naval developments. The emphasis on consensualism offers a Whiggish teleology that is seductive, yet also open to empirical qualification and methodological debate.[30]

Differences between outwardly similar European battle fleets do, nevertheless, raise the question of the relationship between efficient naval performance and specific types of society and political system: if the weapons systems are similar, how far are differences in effectiveness due to the latter? However, it is difficult to measure efficiency and effectiveness. For example, in the eighteenth century the British navy was often not very efficient in the early stages of wars, but in each conflict, after a few years, became much more effective: thus it was more successful in 1782 than in 1778. A similar pattern was true of the British army, with the additional variation that Britain took part in three linked sets of wars (1689–97 and 1702–13, 1743–8 and 1756–63, and 1793–1802 and 1803–15) and, for each set, was more effective in the second conflict.

Rodger's valuable work highlighted the problems posed by any account of military organization that neglects the naval dimension. His published study closes for the present at 1649, but the argument can be taken forward chronologically. Lynn's account of nineteenth-century land warfare usefully focuses on the opportunities and problems posed by the far larger European armies of the period, and takes numbers to be a crucial issue in military organization.[31] The force–number ratio was very different at sea, increasingly so as the force-projection (and cost) of each warship increased with successive advances in gunpowder technology. Whereas galleys had been *primarily* designed to carry troops, rigged warships were built and used far more in order to fight one another. This had obvious implications in terms of their ship-killing requirements: for armament, supply and tactics. In the eighteenth century, individual warships had a greater artillery power than entire field armies. Until the age of air power and space, fleets were the most complex and costly public artifacts of their societies.

If military organization is, in part, understood as an exercise in the control, deployment and sustaining of force, then the situation at sea was obviously different to that on land. The chronology of organizational development was not that outlined for armies by Lynn because, in European navies, professionalization, specialization, machinization and state control all became standard in the sixteenth and, more consistently, the seventeenth centuries.

The technical specialization seen by Lynn as crucial to the situation in the modern Western army began earlier at sea.[32] Furthermore, the situation has not, in essence, changed since: there has been no equivalent at sea to the modern trend towards effective guerrilla warfare, although piracy has revived of late in the Caribbean and off South-East Asia.

Recruitment was an important aspect of naval capability, with, for example, systems of forcible enlistment that mirrored the situation on land, but these generally ended earlier at sea, and there was a greater role for voluntary service, in part because of the lesser need for manpower at sea. In the case of organization, it is also necessary to consider how best to discuss the cohesion and state control of the officer class. They were greater at sea than on land, in large part because military entrepreneurship played a smaller role at sea, while greater professionalism was expected there, and was indeed necessary. This was true of sailing skills, fire control, logistics, and the maintenance of fighting quality and unit cohesion. Fighting instructions and line tactics instilled discipline.[33] Merchantmen ceased to appear in the line of battle of European fleets in the late seventeenth century. The use of East India Company ships as warships in the seventeenth to nineteenth centuries can be regarded as an aspect of military entrepreneurship, although the Companies were in effect branches of European governments.[34]

There was also a difference in the nature of permanent forces between European navies and armies, especially from the sixteenth century. Naval efficiency had to be measured in the ability to create fighting teams for existing ships once the war had begun. The permanent navies were mainly ships and officers, with relatively few sailors. In contrast, the permanent armies were organizational structures of officers and men to which a steady flow of new men had to be recruited.

Naval power – principally Western, although also, in the fifteenth and sixteenth centuries, Chinese and Turkish – dramatically increased the interaction of different parts of the world from the fifteenth century. As before, however, military success involved political incorporation as well as technological strength.[35] This incorporation was an important aspect of military organization, and the successful allocation of the burdens of supporting the military was crucial to the practice and success of incorporation.

The ease of the process of raising men, supplies and money was significant to the harmony of political entities, and thus to the effectiveness of their military forces. Organization has to be understood as political as much as administrative, and indeed, this politicality was more important. Rulers lacking political support found it difficult to sustain campaigns and maintain military organization. Armies could not be recruited, or, once assembled, had to concentrate on the search for supplies, or even mutinied.

The use of agencies and individuals outside the control of the state to raise and control troops and warships was so widespread that the relationship could not be seen simply as devolved administration.[36] This point lessens the contrast that can be made between 'medieval' warfare based on social institutions and structures, and an 'early-modern' system based on permanent organizations maintained and managed by the state. The notion of war and the military as moving from a social matrix – most obviously feudalism – to a political context – states in a states system – is too sweeping.

Between 1490 and 1700, professionalization and the rise of standing (permanent) European forces on land and sea created problems of political and military organizational demand. Structures had to be created, co-operative practices devised, within the context of the societies of the period. It is unclear how far these created a self-sustaining dynamic for change, in an action–reaction cycle or synergy, or how far there was a limit of effectiveness that inhibited the creation of a serious capability gap with forces, both European and non-European, that lacked such development. This is an important issue, given modern emphasis on organizational factors, such as drill and discipline, in the rise of the Western military.[37]

Another important aspect of change and professionalization was provided by the development of an officer corps responsive to new weaponry, tactics and systems, and increasingly trained, at least in part, in a formal fashion, with an emphasis on specific skills that could not be gained in combat conditions. Although practices such as purchase limited state control (or rather reflected the nature of the 'state'),[38] officership was a form of hierarchy under the control of the sovereign. However, most officers came from the social elite, the landed nobility and, at sea, the mercantile oligarchy. An absence of sustained social mobility at the level of military command was an important aspect of organization, and a constraint on its flexibility. It reflected more widespread social problems with the recruitment of talent. Rulers found their ability to select commanders gravely limited. For example, both Louis XIII and Louis XIV of France had to be very careful in the allocation of command among *'les grands'*.[39] The situation in France did not change until the French Revolution brought dramatically new opportunities for the talented.

European forces were not the only ones to contain permanent units and to be characterized by professionalism, but the degree of development in this direction in different parts of the world cannot be easily compared, because of the lack of accurate measures, and indeed definitions. Furthermore, it is necessary to consider how best to weight the respective importance of peacetime forces and larger wartime establishments. Despite these problems, it is important to note the role of permanent units and professionalism elsewhere, including in China, India, Persia, Turkey and Africa.[40]

In accounts of global military history, the 'early-modern' period is generally presented in terms of a European military revolution defined by the successful use of gunpowder weaponry on land and sea. This is then followed by an onset of late modernity, either in terms of a greater politicization and resource allocation, and an alleged rise in determination in war and combat, from the French Revolution, which began in 1789, or in terms of the industrialization of war in the nineteenth century, or of both. Such a chronology, however, is limited as an account of European development, and flawed on the global scale, due to its failure to heed change elsewhere.

In searching for periodization, it is instead important and useful to abandon a Eurocentric chronology and causation. It is possible to suggest a period in addition to those given above, that of 1700–1850, the age before the triumph of the West. It closes in the mid-nineteenth century, when the impact of the West and Westernization first greatly affected areas where they had hitherto had limited effect: Japan, China, South-East Asia, New Zealand, inland Africa, and western interior North America.

The period begins in about 1700, in order to divide the 'early-modern' age. This distinguishes the period from that of the initial expansion of the 'gunpowder empires'. It also focuses on the impact of flintlocks and bayonets, which were important, together, or largely the first, in India, West Africa, Europe and North America.[41] Furthermore, what has been termed the seventeenth-century general crisis that affected so much of the world's economy, with accompanying socio-political strains,[42] was succeeded in the eighteenth century by a period of widespread demographic and economic growth and increased internal stability. This ensured that more resources were available for the military and for war.

Attention in 1700–1850 generally focuses on war within Europe, especially under the command of Charles XII of Sweden, Peter the Great of Russia, the Duke of Marlborough, Frederick the Great of Prussia and Napoleon. However, conflict within Europe was less important in displaying and increasing general European military capability than the projection of European power overseas, a projection achieved in a largely pre-industrial world. To this end, it was the organizational capacity of the Atlantic European societies that was remarkable. The Duke of Newcastle, the First Lord of the British Treasury, claimed in 1758: 'We have fleets and armies in the four quarters of the world and hitherto they are victorious everywhere. We have raised and shall raise more money this year than ever was known in the memory of man, and hitherto at 3½%.'[43] Such a capability was very important in the British conflict with France.

Warships themselves were the products of an international procurement system, and of what were, by the standards of the age, massive and complex manufacturing systems. Their supply was also a major undertaking, as was their maintenance.[44] Neither was effortless, and any reference to the sophistication of naval organization must pay attention to the continual effort that was involved and the problems of supplies.

This can be glimpsed from the surviving correspondence, for example that of Mithon, the Intendant de la Marine at the French Mediterranean base of Toulon, as he sought to surmount difficulties in the wood supply in late 1733 and thus prepare the fleet for the 1734 campaign.[45] France attacked Austria in late 1733, in the War of the Polish Succession. The following year, a French agent commented on the leading Spanish naval base at Cadiz: Spain was also at war with Austria. He noted Spanish efforts to develop a major navy, but reported on a lack of adequate shipworkers and systems.[46] It is necessary to consider such sources alongside the more synoptic, problem-free accounts of organizational development.

Naval supply and maintenance required global networks of bases if the navies were to be able to secure the desired military and political objectives.[47] Thus, the French in the Indian Ocean depended on Mauritius and Réunion, the British on Bombay and Madras, the Portuguese on Goa, and the Dutch on Negapatam. When, in the 1780s, the British considered the creation of a new base on the Bay of Bengal, they acquired and processed knowledge about needs and options in a systematic fashion and benefited from an organized process of decision-making.

The globalization of European power was not solely a matter of naval strength and organization. As discussed in Chapter One, the creation of powerful syncretic Western-native forces, especially from the 1740s by the French, and then the British in India, was also important. A different process occurred in the New World. There, the European military tradition was fractured with the creation from the 1770s of independent forces, beginning with the federal and state, army and naval forces of the American colonies that rebelled against the British Crown in 1775 and declared independence the following year.

The organizational culture and practices of these forces arose essentially from political circumstances. Thus, in the USA there was an emphasis on volunteerism, civilian control and limited size, for both army and navy, that reflected the politics and culture of the new state. This could be seen in Thomas Jefferson's preference for gunboats rather than ships of the line. Nevertheless, there was a professional military in the USA, and not simply a militia. This provided the basis for the subsequent expansion of the American military, not least because a trained officer corps existed that could organize the expansion of manpower necessary when the USA went to war.[48] The USA was not only independent, but also a sufficiently important part of the European world to ensure that the latter had to be reconceptualized as the West. The political culture of the independent states in Latin America that won independence in the nineteenth century from the Spanish and Portuguese empires was different to that in the USA, and this very much affected the organization and character of the resulting armed forces.

Within Europe, there was also a process of combination in raising wartime forces. Armies were largely raised among the subjects of individual rulers, but foreign troops, indeed units, were also recruited, and there was, in addition, a process of amalgamation in which alliance armies were built up.[49] Furthermore, recruiting, in some cases forcible, extended to foreign territories. The Prussians were especially guilty of this process, forcibly raising troops for example in Mecklenburg[50] and Saxony. The creation of coalition forces by amalgamation could involve subsidies,[51] and could also be motivated by operational factors, specifically the recruitment of light cavalry from peoples only loosely incorporated into the state, such as Cossacks for Russia and Crimean Tatars for Turkey. This continued a long-established process seen, for example, with the recruitment of Swiss pikemen to provide 'heavy infantry' in the fifteenth and early sixteenth centuries. Pikemen offered flexibility, both offensive shock and protection for musketeers.

Outside Europe in 1700–1850, expanding powers, for at least part of the period, included China, Burma, Siam, Vietnam, the Afghans, Mysore, the Marathas, the Sikhs, Egypt, Shaka's Zulu kingdom, and the Sokoto Fulani kingdom in West Africa. The organizational character of these forces varied greatly, in large part because of the very different nature of the states involved. The contrast between the methodical campaigns of Chinese forces in Tibet in the 1710s and Zungaria in the 1750s,[52] and the more inchoate political structures, less fixed military organization, and more fluid fighting methods of the Marathas,[53] can be repeated throughout the non-European world. This contrast does not imply that the Marathas were less effective. It is important not to read back from their defeats at the hands of the British in the first quarter of the nineteenth century, and to assume that these defeats and the absence of a strong unitary Maratha state implied a failure of system.[54]

A notion of different and distinctive European and non-European military organizations, and of their related effectiveness, is visually encoded in the art and imagery of (European) empire. It is of the 'thin red line', an outnumbered and stationary European force, drawn up in a geometric fashion, and ready to fire. This force is being charged by a disorganized mass of infantry or cavalry, a mass lacking uniform, formation and discipline.[55] This image can be variously presented – the Europeans, for example, can be in squares rather than lines – but it is designed to suggest the potency of discipline and the superiority of form.

Such an image is central to a teleology of military organization, a notion that organization entails a certain type of order, from which success flows. As an account of the imperial campaigns of the second half of the nineteenth century and of European success, such an image is less than complete, and is, in some respects seriously misleading. The error is even more pronounced prior to the mid-nineteenth century. European forces won major victories at Plassey (1757) and the Pyramids (1798), but not on the Pruth (1711). The organization of forces on the battlefield was only one element in combat; some non-European forces had sophisticated organizational structures, both on campaign and in battle; and European armies themselves frequently did not fulfil the image of poised, coiled power.

This can also be seen by considering the nature of French Revolutionary and Napoleonic battle. These battles were traditionally presented by British historians as an object lesson in the superiority of disciplined organization. The customary view of Wellington's tactics is that his infantry, drawn up in line, stopped oncoming French columns by defensive fire alone; in short, that an organization geared to linear formations was most successful on the battlefield. By extension, the history of successful military organisation in the nineteenth century is thus, in part, an extension of similar formations and practices of control by the Europeans to other parts of the world, as well as emulation by local powers,[56] although, in Europe, such formations were abandoned later in the century, as they represented easy targets for opposing fire, both infantry and artillery.

However, it has been shown that the interaction of organization, discipline and tactics on the Wellingtonian battlefield was different to the conventional view. Wellington's favoured tactic was for his infantry to fire a single volley, give a mighty cheer, and then charge.[57] The key was not firepower alone, but a mixture of fire and shock. This is less different from the situation elsewhere in the world than may be appreciated. In addition, it focuses attention on the role of morale, which is particularly important in shock tactics (although also obviously where there is reliance upon firepower). The rethinking of the Wellingtonian battlefield, for which sources are relatively plentiful, although not without difficulties,[58] poses the question of how far it would be profitable to rethink other battles for which sources are less good. This is important, because assessments of tactical effectiveness frequently rest on accounts that have not been adequately probed.

Shock tactics were, and are, not simply a matter of an undisciplined assault in which organization plays scant role. This can be seen by considering the columnar tactics of European forces in the period 1792–1815. They can be presented as the organizational consequence of the *levée en masse*, the addition by the French Revolutionaries of large numbers of poorly trained conscripts to the army.[59] Yet columns could also be employed on the defensive, a deployment on the battlefield that required a more controlled organization. The formation was an appropriate one in a variety of different ways. First, it was an obvious formation for troops stationed in reserve. Thus, infantry brigades and divisions stationed in the second line would almost always have been in column, this being the best formation for rapid movements, whether it was to plug a gap, to launch a counter-attack or to reinforce an offensive. However, columns were of use even for brigades and divisions stationed in the front line. It was possible to intermix lines and columns in a sort of static version of *l'ordre mixte*, or indeed to have many units in column, as with the Russians opposing Napoleon at Borodino in 1812.

Far, therefore, from presenting columns as a product of a relatively simplistic military organization, a regression from the professionalism, training and sophisticated battlefield manoeuvres of the Frederician army (the military system modelled on the army of Frederick II, the Great of Prussia, 1740–86), it is possible to argue that they were not only effective but a development on what had come before. A sudden onslaught by a line of columns on an attacking force, and particularly an attacking force that had been shot up and become somewhat 'blown' and disorganized, was likely to have been pretty devastating. It was possible to change deployment if advancing in column more easily than if attacking in line. Furthermore, columns could still fire, while they could also be placed side by side to present a continuous front. Columns gave a defender weight, the capacity for local offensive action and solidity, while troops deployed in line came under enormous psychological strain when under attack by columns.[60]

A brief consideration of what actually happened in combat thus reveals the danger of presenting one form of tactical deployment as necessarily weaker than another; a point that is more generally valid, and for sea as well as land operations. For example, seventeenth- and eighteenth-century naval tactics, with their emphasis on parallel lines of opposing vessels firing broadsides, were more intelligent, given the capability of the ships, than subsequent criticism might suggest.[61]

Caution in the assessment of tactical deployment is important in any discussion of organizational structure, because such structures were clearly related to tactical deployment, although the extent to which there was a causal correspondence varied. Another helpful caveat is provided by sources that reveal the limitations of effective diffusion of tactics and weaponry, both within Europe and further afield.

In addition, it is not easy to find any clear correspondence between one type of political system, one set of social and economic preconditions, and a certain level of military and naval efficiency. Organization matters: permanent armed forces can act and react quicker than improvised forces, while tactics which require training, teamwork, trust and solidarity within the fighting unit also favour units that are permanent. Armed forces with conscripts may or may not be successful in creating such units; in large part, this is a matter of military culture and the parameters set by political expectations, rather than flowing necessarily from a particular social or political system. The numbers and enthusiasm of the French Revolutionary forces in the wars that began in 1792 would have counted for far less without a military structure that allowed them to be made use of: divisions, columnar tactics and improved use of light infantry have to be seen, in part, in organizational terms. The divisional (and later corps) system allowed the French to deploy far larger numbers of troops in any given theatre of operations than would otherwise have been the case, and to move them around more quickly and effectively once they had arrived. Equally, better artillery (and artillery tactics), the use of columns, and the co-ordination of close-order and open-order tactics enabled the French to get round the deficiencies of their forces on the battlefield.

The development of the corps system was a good example of the enhancement of military strength by organizational developments. It allowed for much greater flexibility in both attack and defence, and enabled the different arms to be used more effectively. In the 1790s, the French division had been a force of all arms, for example two brigades of infantry, one of cavalry and a battery of guns. This, however, meant that the artillery and cavalry were split up into 'penny packets' and deprived of much striking power. Under the corps system, by contrast, both cavalry and artillery tended to be controlled at a higher level, and even grouped into separate formations altogether, for example artillery reserve and cavalry corps. The result was another 'multiplier', and one used with great effect in the 1800s by Napoleon. More detailed organizational factors were also important. The French were not the first to employ light infantry in a largely separate capacity on the battlefield, but they integrated close-order and open-order infantry so that every infantry battalion had the capacity to deploy its own skirmishers.[62]

Permanent forces are generally seen as crucial to military development, but they may also breed inefficiency and conservatism. Culture matters as much or more than function. The incentive structure matters: if officers are selected, promoted and otherwise rewarded because of proven efficiency, it may pay to take risks and work hard with the training of soldiers or sailors. Societies that are able to create (and finance) permanent forces (not necessarily the largest, but the *best*), and to create efficient incentive structures are likely to have the best armed forces. Such forces and structures have to be sustained. This posed challenges, not simply in maintaining military effectiveness, but also in sustaining related social practices and political policies.[63]

CONCLUSIONS

Any emphasis on a different analysis for military organization on land from the situation at sea is potentially limited by the role of amphibious operations,[64] which became increasingly important in the nineteenth century, and still more from 1942; but that does not challenge the essential distinction. This invites questions about the appropriateness of particular accounts that encompass land and sea and present them as in parallel, such as that of an early-modern military revolution, which is discussed in Chapters Four and Five. The emphasis on firepower in this theory has served to encompass both, but this is limited as an account of operational requirements in battle, let alone other aspects of capability and organization.

In the case of 'modern' (i.e., specialized) navies, the appropriate organizational model is state-driven and state-centred to a degree that was only later true for armies. However, it was, and is still, reasonable to ask how far the general patterns of political and social behaviour affected organizational practice and structure in the armed forces. At the basic level, there was no role for operational consent on the part of sailors in state forces, no more than there was for soldiers. An element of contractualism might pertain for service conditions, especially in terms of payment and the provision of food,[65] but this was lessened in Europe from the mid-seventeenth century as governmental control over the military increased and the provision of services improved. Logistics lessened liberty.

Furthermore, in state forces, contractualism on the part of soldiers and sailors did not extend to operations. Drill required subordination. In 1792, Charles Jenkinson watched a Prussian army review: 'the celerity and precision with which all their movements are performed, are inconceivable to those who have not seen them. Every operation they go through is mechanical.'[66] The French envoy in London was more caustic. He referred to new hordes of disciplined slaves.[67]

However, it is important not to exaggerate the degree of state control. In the case of Britain, there was a balance between local interests and the requirements of the Crown. For example, militia colonels regarded their regiments as patronage fiefs, immensely valuable to them as county magnates and public men, so that important changes in the militia laws had to be negotiated with them, and even the practice of regular drafts into the army was carefully conducted to protect their interests. Military service by militia and the volunteers was conditional.[68] Alongside work on desertion and resistance to service in France,[69] a picture emerges of societies in which the willingness to serve was contingent, and politics, in the widest sense, was as important as government in generating numbers.

The degree of state direction was still very different to the situation in a range of independent or autonomous forces within the European system, such as mercenaries, irregular troops, privateers and pirates,[70] and was also very different to many, but by no means all, non-European forces. Within Europe, there was an important demise in contractualism in the period 1450–1850, as the nature and power of the state were defined and developed from the decline of feudal practices.[71] Societies that offered different models of socio-military organization and that resisted, such as Highland Scotland, the Tyrol, Corsica and the Cossacks, were subjugated. The Swiss were absorbed into the statist military system as an essentially mercenary component of the *ancien régime* French army. By 1850, the Western state and its expectations and assumptions had reached a high point of authority and power, certainly compared with the situation two centuries earlier.

NOTES

1. PRO WO 33/1512, pp. 7–8.
2. A particularly fine example of the former type is G. Parker, *The Military Revolution. Military Innovation and the Rise of the West, 1500–1800* (2nd edn, Cambridge, 1996), and of the latter approach, N.A.M. Rodger, *The Wooden World: An Anatomy of the Georgian Navy* (1986).
3. For example, see D.W. Marshall, 'The British Military Engineers 1741–1783: A Study of Organization, Social Origin and Cartography' (PhD, Michigan, 1976); Gen d'Hollander, 'Oeuvre Cartographique du Dépot de la Guerre', *Revue Historique des Armées*, 171 (1988), pp. 43–4.
4. See, for example, P. Longworth, *The Cossacks* (1969); W. Bracewell, *The Uskoks of Senj: Piracy, Banditry and Holy War in the Sixteenth Century Adriatic* (Ithaca, NY, 1992).
5. Vincennes, Archives de la Guerre, A1 2687.
6. A good example of recent work in this field is E. Kier, *Imagining War: French and British Military Doctrine Between the Wars* (Princeton, NJ, 1997).
7. Richmond to Grey, 27 April 1782, Durham, University Department of Palaeography, papers of 1st Earl Grey, no. 61.
8. W. Treadgold, *Byzantium and Its Army, 284–1081* (Stanford, CA, 1995), pp. 159–66, 188–98; M.C. Bartusis, *The Late Byzantine Army: Arms and Society, 1204–1453* (Philadelphia, PA, 1992).
9. F. Wakeham, *The Great Enterprise. The Manchu Reconstruction of Imperial Order in Seventeenth-century China* (2 vols, Berkeley, CA, 1985).
10. J.W. Tong, *Disorder Under Heaven: Collective Violence in the Ming Dynasty* (Stanford, CA, 1991); P.A. Kuhn, *Rebellion and Its Enemies in Late Imperial China: Militarization and the Social Structure, 1796–1864* (Cambridge, MA, 1970).
11. B.S. Bachrach, *The Anatomy of a Little War: A Diplomatic and Military History of the Gundovald Affair, 568–586* (Boulder, CO, 1995); J.G. Dawson, *The Origins of Western Warfare: Militarism and Morality in the Ancient World* (Boulder, Colorado, 1996). For an emphasis on the limited value of such texts, see D.A. Neill, 'Ancestral Voices: The Influence of the Ancients on the Military Thought of the Seventeenth and Eighteenth Centuries', *JMH*, 62 (1998), pp. 487–520.

12. T.J. Barfield, *The Perilous Frontier: Nomadic Empires and China, 221 BC to AD 1757* (Oxford, 1989).
13. J. Gommans, 'The Silent Frontier of South Asia, *c.* AD 1100–1800', *Journal of World History*, 9 (1998), pp. 1–23.
14. P.M. Malone, *The Skulking Way of War: Technology and Tactics among the New England Indians* (Lanham, MD, 1991).
15. L.V. Eid, 'The Cardinal Principle of Northeast Woodland Indian War', in W. Cowan (ed.), *Papers of the Thirteenth Alonquian Conference* (Ottawa, 1982), pp. 243–50, and '"A Kind of Running Fight": Indian Battlefield Tactics in the Late Eighteenth Century', *Western Pennsylvania Historical Magazine*, 71 (1988), pp. 147–71.
16. J.A. Lynn, 'The Evolution of Army Style in the Modern West, 800–2000', *International History Review*, 18 (1996), pp. 505–45, quotes, pp. 505–6, 508.
17. See, for example, R. Law, '"Here Is No Resisting the Country": The Realities of Power in Afro–European Relations on the West African "Slave Coast"', *Itinerario*, 18 (1994); A.J. Kuethe, 'The Pacification Campaign on the Riohaca Frontier, 1772–1779', *Hispanic American Historical Review*, 50 (1970), pp. 467–81.
18. M. Howard, *War in European History* (Oxford, 1976).
19. J.A. Lynn, 'The Evolution of Army Style in the Modern West', pp. 510, 521, 525.
20. G.W. Johnson (ed.), *Double Impact: France and Africa in the Age of Imperialism* (Westport, CT, 1985).
21. B.W. Menning, *Bayonets Before Bullets: The Imperial Russian Army, 1861–1914* (Bloomington, IN, 1992); W.C. Fuller, *Strategy and Power in Russia 1600–1914* (New York, 1992).
22. M.E. Mallett and J.R. Hale, *The Military Organization of a Renaissance State: Venice, c. 1400–1619* (Cambridge, 1984), pp. 71–4; Mallett, 'Venice and the War of Ferrara, 1482–84', in D.S. Chambers, C.H. Clough and M.E. Mallett (eds), *War, Culture and Society in Renaissance Venice* (1993), p. 66.
23. L. Woolf, *Inventing Eastern Europe: The Map of Civilization on the Mind of the Enlightenment* (1994).
24. The recent work on non-Western dimensions can be approached through D.M. Peers (ed.), *Warfare and Empires: Contact and Conflict Between European and Non-European Military and Maritime Forces and Cultures* (Aldershot, 1997), and J.M. Black, *War and the World 1450–2000* (New Haven, CT, 1998).
25. J. Gommans, *The Rise of the Indo-Afghan Empire c. 1710–1780* (Leiden, 1995).
26. J.M. Hill, 'Celtic Warfare, 1453–1815', in J.M. Black (ed.), *European Warfare, 1453–1815* (1998). For an emphasis on development of the Scottish military establishment on the same lines as other European armies, and on the influence of Continental European precedents on Highland tactics, G. Phillips,

'In the Shadow of Flodden: Tactics, Technology and Scottish Military Effectiveness, 1515–1550', in *Scottish Historical Review*, 77 (1998), pp. 162–82.

27. See, for example, A. Ayton and J.L. Price, 'The Military Revolution from a Medieval Perspective', in Ayton and J.L. Price (eds), *The Medieval Military Revolution: State, Society and Military Change in Medieval and Early Modern Europe* (1998), p. 15.

28. N.A.M. Rodger, *The Safeguard of the Sea: A Naval History of Britain, Volume I: 660–1649* (1997), pp. 430–4.

29. J.C.K. Daly, *Russian Seapower and 'the Eastern Question', 1827–41* (1991).

30. For a similar approach for the eighteenth century, see J. Brewer, *Sinews of Power: War, Money and the English State, 1688–1783* (1989).

31. J.A. Lynn, 'The Evolution of Army Style in the Modern West', pp. 521–2.

32. C.R. Phillips, *Six Galleons for the King of Spain: Imperial Defence in the Early Seventeenth Century* (Baltimore, MD, 1992); J.R. Bruijn, *The Dutch Navy of the Seventeenth and Eighteenth Centuries* (Columbia, SC, 1993); D. Goodman, *Spanish Naval Power, 1589–1665: Reconstruction and Defeat* (Cambridge, 1996).

33. M.A. Palmer, '"The Soul's Right Hand": Command and Control in the Age of Fighting Sail, 1652–1827', *JMH*, 61 (1997), pp. 679–706.

34. L. Blusse and F. Gaastra (eds), *Companies and Trade: Essays on Overseas Trading Companies During the Ancien Régime* (The Hague, 1981).

35. For an example within Europe, see D. Potter, *War and Government in the French Provinces: Picardy, 1470–1560* (Cambridge, 1993).

36. F. Redlich, *The German Military Enterpriser and His Work Force: A Study of European Economic and Social History* (2 vols, Wiesbaden, 1964–5); D. Kolff, *Naukar, Rajput and Sepoy: The Ethnohistory of the Military Labour Market in Hindustan, 1450–1850* (Cambridge, 1990). For the situation at sea in the European colonial world, see C.C. Goslinga, *The Dutch in the Caribbean and on the Wild Coast, 1580–1680* (Assen, 1971); K.R. Andrews, *Trade, Plunder and Settlement: Maritime Enterprise and the Genesis of the British Empire, 1480–1630* (Cambridge, 1984).

37. D. Showalter, 'Caste, Skill, and Training: The Evolution of Cohesion in European Armies from the Middle Ages to the Sixteenth Century', *JMH*, 57 (1993), pp. 407–30.

38. T. Barker, *Army, Aristocracy, Monarchy: Essays on War, Society and Government in Austria, 1618–1780* (Boulder, CO, 1982); G. Rowlands, 'Louis XIV, Aristocratic Power and the Elite Units of the French Army', *French History*, 13 (1999), pp. 303–31. Victor Amadeus II of Savoy-Piedmont was able to avoid extensive venality by obtaining foreign subsidies, C. Storrs, *War, Diplomacy and the Rise of Savoy, 1690–1720* (Cambridge, 1999), pp. 316–17.

39. D. Parrott, 'Richelieu, the *Grands*, and the French Army', in J. Bergin and L. Brockliss (eds), *Richelieu and His Age* (Oxford, 1992), pp. 135–73; G.

59

Rowlands, *Power, Authority and Army Administration under Louis XIV: The French Crown and the Military Elites in the Era of the Nine Years War* (DPhil., Oxford, 1997). For the wider social politics of military service, P.H. Wilson, 'Social Militarization in Eighteenth-Century Germany', in *German History*', 18 (2000), pp. 1–39.

40. See, for example, W. Irvine, *The Army of the Indian Moghuls: Its Organization and Administration* (Delhi, 1962).

41. See, for example, R. Hellie, 'The Petrine Army: Continuity, Change, and Impact', *Canadian-American Slavic Studies*, 8 (1974), pp. 239–40.

42. G. Parker and L.M. Smith (eds), *The General Crisis of the Seventeenth Century* (1978).

43. Newcastle to Andrew Mitchell, 12 September 1758, BL Add. 6832 fol. 43.

44. R. Morriss, *The Royal Dockyards During the Revolutionary and Napoleonic Wars* (Leicester, 1983).

45. Paris, Archives Nationales, Archives de la Marine, B3 359 fols 139–40, 149, 179–81, 189.

46. D'Orves to Chauvelin, Foreign Minister, 4 September 1734, AE CP Espagne 419 fol. 144.

47. G.D. Inglis, 'The Spanish Naval Shipyard at Havana in the Eighteenth Century' in *New Aspects of Naval History: Selected Papers from the 5th Naval History Symposium* (Baltimore, MD, 1985).

48. S.C. Tucker, *The Jeffersonian Gunboat Navy* (Columbia, SC, 1993); W. Sword, *President Washington's Indian War: The Struggle for the Old Northwest, 1790–1795* (Norman, OK, 1985); R. Kohn, *Eagle and Sword: The Beginnings of the Military Establishment in America* (New York, 1975).

49. D. Chandler, *Marlborough as Military Commander* (1973).

50. Memorandum by Mecklenburg government, winter 1757–8, Paris, Bibliothèque Victor Cousin, Fonds Richelieu, vol. 58 fol. 247.

51. See, for example, C.W. Eldon, *England's Subsidy Policy Towards the Continent During the Seven Years' War* (Philadelphia, PA, 1938); P.G.M. Dickson, *Finance and Government under Maria Theresia 1740–1780* (2 vols, Oxford, 1987), Vol. II, 158–69, 172–3, 391–3, 395; P.H. Wilson, *German Armies: War and German Politics 1648–1806* (1998), pp. 205–9; P.K. Taylor, *Indentured to Liberty: Peasant Life and the Hessian Military State, 1688–1815* (Ithaca, NY, 1994).

52. P.C. Perdue, 'Military Mobilization in Seventeenth and Eighteenth-Century China, Russia, and Mongolia', *Modern Asian Studies*, 30 (1996), pp. 757–93.

53. S. Gordon, *The Marathas 1600–1818* (Cambridge, 1993).

54. For Maratha fighting capability against the British, see R.G.S. Cooper, 'Wellington and the Marathas in 1803', *International History Review*, 11 (1989).

55. See, for example, T. Porterfield, *The Allure of Empire: Art in the Service of French Imperialism 1798–1836* (Princeton, NJ, 1998), pp. 49, 62–4.

56. R.J. Smith, *Mandarins and Mercenaries: 'The Ever-Victorious Army' in Nineteenth Century China* (Millwood, NY, 1978); R.J. Samuels, *Rich Nation, Strong Army: National Security and the Technological Transformation of Japan* (Ithaca, NY, 1994); A. Amanat, *The First of the Universe: Nasir al-Din Shah and the Iranian Monarchy, 1831–1896* (1996).

57. P. Griffith (ed.), *Wellington Commander* (Chichester, 1985).

58. R. Muir, *'So Brilliant a Victory': Wellington at Salamanca* (New Haven, CT, 2001).

59. J.A. Lynn, *The Bayonets of the Republic: Motivation and Tactics in the Army of Revolutionary France* (Urbana, IL, 1984).

60. On the flexibility of columns, see J. Chagniot, *Le Chevalier de Folard, la stratégie de l'incertitude* (Paris, 1997).

61. M.A. Palmer, 'The "Military Revolution" Afloat: The Era of the Anglo–Dutch Wars and the Transition to Modern Warfare at Sea', *War in History*, 4 (1997), pp. 123–49.

62. D. Chandler, *The Campaigns of Napoleon* (1966); G.E. Rothenburg, *The Art of Warfare in the Age of Napoleon* (Bloomington, IN, 1978); G. Nafziger, *Imperial Bayonets: Tactics of the Napoleonic Battery, Battalion and Brigade as Found in Contemporary Regulations* (1996); B. Nosworthy, *Battle Tactics of Napoleon and his Enemies* (1996); R. Muir, *Tactics and the Experience of Battle in the Age of Napoleon* (New Haven, CT, 1998).

63. See, for Russia in 1726 after the death of Peter the Great the previous year: *Sbornik Imperatorskago Russkago Istoricheskago Obshchestra* (148 vols, St Petersburg, 1867–1916), Vol. LXIV, 241–2.

64. See, in particular, R. Harding, *Amphibious Warfare in the Eighteenth Century: The British Expedition to the West Indies, 1740–1742* (Woodbridge, 1991), pp. 150–97.

65. G. Parker, *The Grand Strategy of Philip II* (New Haven, CT, 1998), pp. 130, 136.

66. Jenkinson to Lord Hawkesbury, 25 July 1792, Bod. Bland Burges papers 37 fol. 62.

67. Chauvelin to Lebrun, Foreign Minister, 9 October 1792, AE CP. Angleterre 582 fol. 318.

68. J. Cookson, *The British Armed Nation, 1793–1815* (Oxford, 1997).

69. A. Forrest, *Conscripts and Deserters: The Army and French Society During the Revolution and Empire* (Oxford, 1989).

70. J.E. Thomson, *Mercenaries, Pirates, and Sovereigns: State building and Extra-territorial Violence in Early Modern Europe* (Princeton, NJ, 1994).

71. B.M. Downing, *The Military Revolution and Political Change: Origins of Democracy and Autocracy in Early Modern Europe* (Princeton, NJ, 1993); B.D. Porter, *War and the Rise of the State: The Military Foundations of Modern Politics* (New York, 1994).

THREE

The Development of Military Organizations 1850–2000

Each individual infantry soldier must be trained to be a self-reliant big game hunter imbued with a deep desire to seek out and kill his quarry.
British Military Mission report on how best to fight Japanese,
April 1944[1]

From the mid-nineteenth century, the world was increasingly under the sway of the West, directly or indirectly. The organization of the Japanese army, in response first to French and then to German models, and of the navy, under the inspiration of the British navy, was a powerful example of this impact.[2] Such emulation, however, was more than a matter of copying a successful military machine. There was also a socio-political dimension that focused, in particular, on the impact of nationalism, but also on other aspects of nineteenth-century 'modernization'.

Conscription was seen in Continental Europe as important to the ideological and political programme of modernization. Although systems of conscription did not require nationalism, they were made more effective by it. Nationalism facilitated conscription without the social bondage of serfdom, because conscription was legitimized by new ideologies.[3] It was intended to transform the old distinction between civilian and military into a common purpose.

Although the inclusive nature of conscription should not be exaggerated, it helped in the militarization of society,[4] and, combined with demographic and economic growth, provided nineteenth-century Western governments with manpower resources such that they did not need to turn to military entrepreneurs, foreign or domestic. Instead, they sought to bolster their strength in war by alliances with other states. The political and ideological changes and increasing cult of professionalism of the nineteenth century also made it less difficult for many states to control their officer corps and to ensure that status within the military was set by government, although the military remained important in influencing policy, and in some states, especially in Latin America, civilian control was limited.[5] Furthermore, the employment of the military to maintain order in several European states, and of the role of the army as a police force in being, was an important part of the social politics and military history of the period. In 1815–1914, this was true of France, Germany and Russia, among other states. Although large-scale internal policing by the military was only episodic in character, it was very important when it occurred, and the likely attitude of the army to such tasks was seen as an important political factor.

There was a heavy emphasis on discipline within armies. Captain Holland, British observer at the German manoeuvres of 1898, commented on: '. . . a system by which losses of smaller units are ignored, provided the main object is gained, the great aim being to train the soldier to carry out his orders regardless of consequences'.[6] Two years earlier, Capt Birkbeck commented: 'It is impossible not to be deeply impressed by the smoothness and ease with which the German military machine works . . . a well-trained and thoroughly practical staff. . . . The German army corps is no collection of units hurriedly collected for a time.'[7]

Yet these disciplined military systems were not inflexible, indeed far from it, for it proved relatively easy to introduce innovations and to respond to the pace of technological advance. The latter was seen, for example, in the rapid and almost continual adoption of new types of warship, naval armament and naval tactics in the period 1840–1940 in response to fundamental changes in ship power and design, and in naval potential at sea.

Innovations can also be seen on land. They were noted, for example, by foreign military observers who were very keen to discern signs of change. In 1900, observers noted that volley firing was being abandoned in favour of individual fire.[8] In addition, Belgian manoeuvres displayed the defensive strength of trenches and the lessons being learned from the Boer War then being waged.[9]

As a result of nationalism and the attendant increase in the scale of mobilization of resources, it became more apparent that war was a struggle between societies, rather than simply armies and navies. This poses questions as to how best to define and understand military organization in this period. If society was mobilized for war, as was indeed the case in both World Wars, then large sections of the economy were directly placed under military authority and became part of the organization of the militarized state. Other sections were placed under governmental control and regulated in a fashion held to characterize military organization. The Ministry of Munitions that was created in Britain in the First World War was as much part of the military organization, and as vital, as the artillery it served.[10] Other sections of society were not brought under formal direction, but can be seen as part of the informal organization of a militarized state.

Universal military training and service were expanded in the First World War. Conscription was introduced in Britain (1916), Canada (1917) and the USA (1917).[11] This accentuated a pattern of military organization that was not based on voluntary service. Such a pattern was central to the armed forces of the combatants in both World Wars. The resulting forces were enormous. In June 1941, the Soviet army west of the Urals was 4.9 million strong.

The situation altered after the Second World War, in large part in response to the impact of individualism in Western society. Other factors were naturally involved in the abandonment of conscription, not least cost and the growing sophistication of weaponry, but they would not have been crucial had there not been a major cultural shift away from conscription. This shift is the most important factor in modern military organization, because it has opened up a major contrast between societies that have abandoned conscription and those where it remains normal. The conflation of volunteerism with the changing nature of citizenship and the decline of deference poses challenges to patterns of obedience within the modern army, and places strain on organizational ideology and coherence. This has been a major issue for several decades, most obviously with the USA during the Vietnam War, but also affecting far less liberal societies, for example the Soviet Union in the 1980s and 1990s.

Again, however, it is necessary to avoid any sense of a clear-cut model and an obvious teleology. Thus, the pattern in Britain was one of a hesitant approach towards conscription, even when it appeared necessary, as in the First World War. In 1780, when the British were at war with France, Spain, Mysore and the American rebels, Charles Jenkinson, the Secretary at War, wrote to Lord Amherst, the Commander-in-Chief, that he did not see how the strength of the army could be maintained, but he added:

> I am convinced that any plan of compulsion in a greater context is not only contrary to the nature of the government of this country, but would create riots and disturbances which might require more men for the purpose of preserving the peace, than would be obtained by the plan itself . . . besides, that, men who are procured in this way almost constantly desert, or at best make very indifferent soldiers.[12]

In the West, war in the twentieth century became less frequent, and thus less normal and normative. Instead, wars are increasingly perceived as aberrations best left to professionals. The training that this professionalism entails has become both lengthier and more specialized. This is important for the growing 'differentness' of military service and training, and for the process of specialization and thus distinction within the military. The British army committee that in 1938 recommended the merger of the Royal Tank Corps and the newly mechanized cavalry noted that, in the past, troops had been trained within their own regiments, but that 'this system is impracticable for a corps equipped with armoured fighting vehicles, and it is clear that in future training will be necessary at a depot equipped with suitable vehicles and staffed by technically qualified instructors'.[13]

Specialization of function is not, however, the major reason why in the West there has been a growing reluctance to employ force in domestic contexts. Instead, political culture is the issue. Governments prefer to rely on the police to maintain internal order, and the use of troops in labour disputes became less common than earlier in the twentieth century. In Britain, troops were deployed during dock strikes in the late 1940s, but not against strikes by coal miners in 1972, 1974 and 1984–5 that posed a serious challenge to the power and authority of the government.

Britain phased out conscription in 1957–63, and the USA moved in 1973 to an all-volunteer military[14] that reduced popular identification with the forces. It is now unclear whether a major sustained conflict in which such states were attacked would lead to a form of mass mobilization. At present, that seems unlikely, for both political and military reasons, but were it to occur, another world war might lead to a different situation, with mobilization designed to engender and sustain activism as much as to provide military manpower. The abandonment of conscription is an aspect of the manner in which the size and purpose of the military are set by political factors and subject to political debate.

Shifting concepts of active citizenship have also played a major role in the changing gender politics of military obligation and service. For most of history, military organization has been essentially a matter of male activity, although women have had to bear many of the resulting costs. This situation is currently changing, in part because the premium on human strength is now far less, but largely, due to ideological rather than functional considerations: a response to the role of women in modern Western society and the need for institutions to adapt accordingly. Any future typology of military organization will need to address this point, and may also need to re-examine issues of ethnicity and incorporation in a modern context.

Control over the military has remained an important issue. The professionalization of armed forces has made it more so by increasing their sense of separateness and their coherence. It would, of course, be misleading to imply that challenges to political control from within the military are only recent. In March 1793, Dumouriez, commander of the French Army of the North, concluded agreements with the Austrian commanders that were intended to allow him to suppress the revolution in Paris.[15] Pressure from military leaderships was frequently blamed for aggressive state policies in the nineteenth century. Contested state direction on the part of all or much of the officer corps became more common in the twentieth century, with the discrediting or overthrow of traditional regimes and patterns of control, and with the increased frequency of domestic instability. This encouraged a rise in the frequency of successful and attempted military coups, as, for example, in Turkey after the First World War, Germany in 1920, Persia (Iran) in 1921 and 1926, Prussia in 1932, Spain in 1936, Yugoslavia in 1941, Germany in 1944, Iraq in 1958, France in 1961, South Vietnam and Syria in 1963, Greece in 1967, Portugal in 1974, Spain in 1981, and Russia in 1991. The list of coups and attempted coups could be readily extended.

In areas where democratic culture is weak, especially the notion of losing power through peaceful means, the military frequently plays a central role in attempts to arrange changes in government.[16] In Pakistan, the military under General Ayub Khan seized power in 1958, creating a government only replaced by another coup, by General Yahaya Khan, in 1969. Defeat by India in 1971 led to the return to civilian rule, but in 1977 another coup put General Zia ul-Haq in control. He retained it until killed in 1988 in an air crash, probably caused by a bomb. The army seized power again in 1999. This was not the sum total of military intervention in Pakistan, as army support for 'constitutional coups' was also crucial, for example in 1993.

Such coups need to be included in any analysis of military organization, because they pose important questions about the nature of governmental control and the character of the state matrix of military power. Even where the military did not formally take over the state, they could still seize control of policy, as in Japan in the 1930s.

An account of military organization as a product of politics is not intended to serve as a demilitarization of military history, but it underlines the point that the notion and understanding of fitness for purpose are essentially set by those who control the military. In some situations, this control is vested in the military. That is the case when the political and military leaderships are similar. This is true of military dictatorships, both modern ones and their historical progenitors, such as those Roman regimes presided over by a general who had seized power, for example Vespasian. Napoleon I was a later example of the same process, and it is not surprising that he plundered the imagery of imperial Rome for motifs. In some societies, such as those of feudal Europe, it may not be helpful to think of separate military and political classifications of leadership.

On the whole, in the twentieth century, military leaders had only a limited success in determining policy. In wartime, generals were sometimes able to gain control of the definition of what was militarily necessary, both in terms of means and objectives. Civilian leaders sought, with only limited success, to resist this process in the First World War. However, in dictatorial regimes, such as those of the Soviet Union and Germany during the Second World War or modern Iraq, the generals were only heeded if their views accorded with those of the dictator.

The Communists' People's Liberation Army became the Chinese army in 1949. Party control ensured that warlordism was both far less than earlier in the century and differently expressed. In turn, the army was used to support the domestic policies of the government. In 1968–9, in response to the growing volatility of the Cultural Revolution, the army restored order and enforced Mao Zedong's control, and this military role was only slackened in 1971. In 1989, again under governmental and party control, the army suppressed demonstrations in Beijing.[17] In democracies, generals have also been subject to political direction, although with less bloody consequences. MacArthur was conspicuously sacked by Truman in 1951.

An important aspect of the shift in the state matrix of military power was provided by greater signs of independence on the part of ordinary soldiers and sailors in the twentieth century. There had been important episodes of discontent over conditions and political activism in the 1790s and nineteenth century, but they became more frequent in the twentieth. After serious disaffection in the Russian army in 1905–7, in 1917–19 military disaffection helped to overthrow the governments (and military authority practices) of

Russia, Austria and Germany, and nearly had that effect in France too.[18] Indeed, the response among the troops to the losses in the Nivelle offensive in 1917 ensured that, for the remainder of that year and in early 1918, the French army refrained from offensive operations. The ability of Britain to intervene in the Russian Civil War in 1918–19 was lessened by signs of disaffection in the military, particularly a widespread demand for demobilization. The consequence of such disaffection encouraged both governmental anxiety about the military, and concern within the latter about the impact of political developments and agitation.

Disaffection on the part of ordinary soldiers played a smaller role in the Second World War. Although it contributed to the collapse of Italy's war effort in 1943, the German and Japanese military did not suffer a comparable collapse. In the postwar world, the issue of military morale was important in the wars of decolonization, in the conflicts of the Cold War, especially Vietnam and Afghanistan, and in localized wars, such as the Middle East and over the Falklands in 1982. The poor morale of the Argentinian defenders helped the British when they retook the Falklands.

In the 1990s, there was considerable uncertainty within the West about military and, particularly, civilian willingness to accept combatant losses in war. This very much differs from the situation in authoritarian societies, such as Iraq and Burma. Israel is an example of a state where losses are very sensitive but there is a greater willingness to fight than in Western societies.

These contrasts can be regarded as an important aspect of modern military organization, for such organization is in large part set by wider social, cultural and political parameters. Indeed, the decline of the coercive power of government forms an ironic counterpoint to claims that, thanks to a contemporary RMA (Revolution in Military Affairs), war will be more effective than ever before. Technology and information are not, however, the sole measures of effectiveness, no more than they were in the past.

If the modern situation invites a contrasting of Western and non-Western circumstances and developments (not that authoritarianism and the non-West are coterminous, nor that there is any unity of experience among the West or the non-West), this suggests that it may be necessary to look back and to shift the emphasis from the diffusion of Western machines, methods and models to the resilience of Western/non-Western differences, a theme probed in the subsequent chapters.

Weapons and tactics were rather easily spread if they were successful, but efficient military performance was much more difficult to create or imitate. Efficiency during combat and difficult operational movements seems invariably to have been connected with good officers at all levels (including NCOs). They had to be dedicated to their job, and to put an emphasis on recruitment and training. This was the case not only in state forces, but also in mercenary units. Unit cohesion was greatly enhanced if soldiers and sailors understood that officers who might be hard drill-masters gave them the best insurance against defeat and death in combat. Officers (and rulers) had to be careful in recruiting new officers, and giving them proper training and reproducing the proper attitude, in order to maintain organizational efficiency, but these notions were culturally conditioned.

The processes by which commanders and officers were chosen were also culturally conditioned. George Thomas, an Irish adventurer who in 1797–1802 gained, by conquest, an independent principality north-west of Delhi, recorded of the Rajah of Jaipur:

. . . one thing only tends to mark the deficiency in the otherwise sound policy of the Prince, which is the appointment of people of inferior rank to high commands in his armies. Naturally high spirited and haughty, a Rajput is of all other people averse to serve under a person whom he deems inferior to himself. Hence arises the impolicy of the Rajah who has of late years introduced into commands in his army slaves, menial servants, mechanics and others, who exclusive of want of talents, and abilities, cannot be supposed to possess that independency of spirit which alone excites to the performance of great actions.

Some will assert, however, in justification of the Rajah's measures in this instance, that the difference of cast and condition makes none in respect to courage, which Mr. Thomas observes may hold good with respect to European troops, or Indian troops disciplined and conducted by European officers in which instance they may be considered as a machine actuated and animated by the voice of the commander; but in an irregular army where discipline never obtained little can be expected from chiefs who in their actions are not stimulated by a sense of personal honour. The Rajputs therefore . . . deem their commanders a disgrace and consequently are indifferent in their obedience and acknowledgement of authority.[19]

It would be mistaken to imagine that caste was not also a factor in European forces.

Morale and organization were, and are, closely linked. Morale requires confidence in colleagues, superiors and purpose. This was true not only of Western forces, but also of their non-Western counterparts, although the consequences in terms of fighting method might be very different. The perseverance and loyalty of the common soldiers of the early-modern Ottoman army, as of their opponents, owed much to incentives, bonuses, booty and ceremonials. The relationship between such similarities (and differences) and ideological factors is unclear.[20]

Any stress on cultural parameters in terms of different understandings of victory, loss and suffering complements those of differences in hierarchy, discipline and drill between powers, especially but not only Western and non-Western. However, the different understandings of victory, loss and suffering are more important as activators of asymmetry. This was seen with the American intervention in Vietnam.[21] If effectiveness is, in part, a matter of the setting of realizable goals and the response to these goals,[22] then the processes by which goals are defined and disseminated can be seen as a conflation of the narrower and the wider definitions of military organization discussed in the second paragraph of the previous chapter.

This conflation should be approached in terms of a cultural relativism that challenges any easy notion of a Europe-centred diffusion. For example, the Manchu Banner system might be seen as the most effective organizational development of the seventeenth century (if such global rankings are not dismissed as unhelpful). Alternatively, it can be seen as evidence of organizational effectiveness and change in a non-Western context.

To return to the question of whether it is possible to offer a taxonomy and analysis of change on the global scale, the suggestion here is that a situation of differing systems, linked to the presence or absence of strong central governments, was replaced in the nineteenth century by the triumph of Western models of governmental and military organization, whether due to colonial conquest by Western powers, or to developments outside the scope of Western control, but reflecting (at least in part) Western models. Modernity, defined in a Western fashion and as a Westernizing project, emerged in large part through military forces operating under Western systems of control and discipline.[23]

However, similarities in army and naval organizational structures (and weaponry) did not imply a convergence in terms of the cultural suppositions affecting war, especially understandings of victory, loss and suffering. This is important, because military organization has, in part, to be understood as intended to achieve particular goals; the functionality was set by an intentionality that was culturally contingent. Thus, just as it is appropriate to query an attempt to analyse military technology without thinking in terms of particular contexts, so the same is true of military organization.

One particular dimension that changes is that of the continuum between specialization of military service and mass participation. This can, in part, be approached by analyses of systems and ethos of volunteerism and conscription. The continuum affects the nature of military organization, even if it does not affect its institutionalization in terms of unit hierarchy. Thus, for example, systems reliant on conscription face different problems in terms of the incorporation of soldiers and officers, in particular with reference to training and discipline, to those confronting militaries reliant on voluntary service. Such differences lead to an analytical emphasis on military cultures understood as representing social and political practices, and again, the emphasis is on diversity. For example, the ethos of a multi-ethnic conscript army where the ethnicities are not treated equally – a frequent situation in military history – are different to those of more homogenous forces. Such differences have a temporal as well as a geographical component, but it would be misleading to see a unitary pattern of temporal change, particularly if a global perspective is adopted. Thus, the European colonial empires made different use of native troops, both for integrative and for repressive ends.

Furthermore, it would be misleading to adopt a teleology. Mass conscript-based forces, for example, can be seen as the product of specific circumstances, rather than a stage in military progress. As such, the inherent 'efficiency' in operational terms of such forces can be questioned: it has to be related to the context. The very notion of military progress can be questioned from this perspective. In 1792, the British Resident at the court of the Maratha Peshwa at Poona (Pune) was given a defence of conventional Maratha methods of warfare:

Their prosperity absolutely depended on their desultoriness, that to alter their mode would require ages and they had already experienced misfortunes by untimely attempts to change the national system particularly in the loss of the battle of Panipat [1761] where Sudaba attempting to substitute a more regular style to their predatory one had been totally defeated.[24]

Scepticism about the necessary effectiveness and progressive role of particular types of forces also emerged from a consideration of 'revolutionary' armies. The misleading nature of the myths associated with such forces emerged from S.P. Mackenzie's *Revolutionary Armies in the Modern Era: A Revisionist Approach* (1997). Instead of triumphs of the will, Mackenzie emphasized the extent to which those who fought often did so for a variety of non-altruistic motives, and also the degree to which chance and the actions of the enemy were crucial to outcomes. Thus, the New Model Army of the Parliamentarians in the English civil wars of the 1640s was presented as far from revolutionary: Mackenzie argued that an insufficiently critical approach has been adopted towards the accounts of the army's chaplains, a reminder of the need to re-examine the empirical basis of assessments of capability already noted in the consideration of Wellingtonian tactics. The role of chance in its decisive victory at Naseby (1645) was emphasized, and Mackenzie suggested that the most distinctive feature of the army was its combat experience, an operational point that could also be made about the effective Scottish army of the early 1640s (see also Chapter Six). Luck and circumstances, not moral status, were seen as crucial in the American War of Independence. The armies of the French First Republic were seen as neither politically motivated nor militarily innovative, and Mackenzie suggested that myths about them had a disastrous impact in France in 1870–1, when the Government of National Defence sought to rely on the numbers and zeal of citizen armies. The heroic image of the Latin American Wars of Liberation was debunked, as was that of the mid-nineteenth-century Chinese Taiping rebels. The pre-modern world view of the sepoys in the Indian Mutiny of 1857–8 was stressed, and the crucial absence of operational cohesiveness traced to the mutiny. Mackenzie closed with an account of the Viet Cong, a force whose revolutionary credentials and mythological strength, he argued, derived, in large part, from its American opponents. He claimed that individual or family hopes and fears were crucial to recruitment, and that the Viet Cong leaders exaggerated the potential of revolutionary spirit and their own offensives.

More generally, there is a need to examine not only 'revolutionary' forces, but also conventional operations from a perspective that is not conditioned by a priori assumptions. A historical perspective is very helpful, as it highlights the variety of past military forms and practices. It can also serve to remind about past discussion of various options. For example, Clausewitz, frequently seen as the analyst of conventional warfare by regular armies, also pressed for the use of militia when Prussia turned against Napoleon in 1813 in a so-called war of liberation, and discussed 'people's war' in his *On War* (1832),[25] although his criticism of Wellington's strategy in 1815 reflected a preference for a decisive battle, rather than Wellington's emphasis on the politico-strategic aims of the alliance.[26]

In his study, Mackenzie underlined the double standard applied to results. Victory for the revolutionary side is seen as a sign of relative moral strength, defeat as the product of betrayal. This raises the more general issue of how best to establish an empirical basis for the analysis of military success – a question which underlines the problem of how to assess development.

Again, there is a tension between the desire to find form and the need to note variations that subvert any clear analytical pattern. There is also the problem of determining whether it is pertinent to note anything distinctive about military organization, or whether it is more appropriate to offer an account in which the military is simply another aspect of political and governmental change, a focus of state activity, but not inherently distinct from other state activities. Thus, for example, the rise of mass reserve forces in Europe in the late nineteenth century can be related as much to nationalism, the creation of national educational systems, and universal male franchises as to anything that is distinctly military: political context, rather than military purpose, was crucial. If this approach is adopted, it would be interesting to consider whether conscription and mass reserve forces were, or are, more valid for areas where there was relatively little conflict, for example Latin America since 1890, and not for those of greater warfulness.

This question can be taken a stage further by asking whether differences in military structures and in the assessment of success can be related to the question of primacy of opponent. Are the ethos of military organizations different if the principal likely foe is domestic, rather than foreign? In the former case, the nature and development of organizations is likely to be greatly affected by the prospect of civil war or insurrection, both at the level of officer corps and that of ordinary soldiery. Thus, conscription may be adopted, abandoned or altered in response to the domestic situation. Such an approach appears particularly appropriate in the case of Latin America,[27] or Mediterranean Europe since 1945.

The categories 'foreign' and 'domestic' can be subdivided. In the latter case, there are important functional and ideological differences between, for example, peasant risings, urban discontent and coups from within the military-political system. In the case of foreign challenges, there are also major differences, not least those arising from the nature of neighbouring powers, the character of alliances, and the range of projection of power.

Thus, for example, France and China were, and are, very different military powers, and contrasts in military role affected and affect expectations and organization. The Chinese military has not been called upon to project its power great distances during the period of military modernization in the last two centuries. In the 1870s and early 1880s, the Chinese preferred to intervene in northern Vietnam against the French by supporting the unofficial Black Flag Army, rather than by sending regular units. In the Second World War, Chinese forces were mostly engaged within China, although about 100,000 men were sent to Burma to resist the Japanese advance in 1942, and fresh forces were committed in 1944. Subsequent to the Chinese Civil War (1945–9), Chinese border conflicts or confrontations with India, the USSR and Vietnam, and massive intervention in the Korean War (1950–3) did not match the long-distance military activity of the USA, Britain, France, and even Cuba. In part, this was a matter of the absence of trans-oceanic relics, such as the Falkland Islands, or post-colonial commitments, as with the French in Africa.

That, however, reflected a central characteristic of non-Western military systems: their limited global range. This is more a matter of political interests, than the range of weapon systems, but these interests are crucial in creating the parameters within which military effectiveness has to be assessed, and thus in stressing the relative nature of such assessments.

In the nineteenth century, the measure of Chinese strength was the ability to resist the Europeans and Japan, not the success of establishing colonies in Africa, the Pacific or even South-East Asia. Similarly, in the post-1949 world, the Chinese have not sought to play a role in Africa or the Middle East akin to the USSR or Cuba. China has a major fleet, and in the 1990s was, at least in terms of numbers, the fourth strongest naval power in the world. In 1964, the Chinese exploded their first atomic bomb – the fifth power in the world to do so – and in 1967 their first hydrogen bomb – the fourth power in the world to do so.[28] Yet the Chinese have not sought a global role to match their military power. They may, however, find that their regional hegemonic pretensions, especially as far as Taiwan, Korea and the South China Sea are concerned, lead them into a more major confrontation than they desire, a confrontation that will pose the intellectual problem that has been largely evaded: namely, how to assess the respective military capability of states that are very different in their political cultures and geopolitical situations.

The respective strength of China and its likeliest international rival in the world, the USA, is clearly a matter not only of weaponry and economic resources, but also of a willingness to inflict and, still more, sustain casualties that is a central facet of political culture. Military historians, with their preference for order, have often emphasized the strength of authoritarian societies, such as China. However, in so far as military capability in the next century may *in part* be a matter of the so-called RMA – the Revolution in Military Affairs that involves paradigm shifts in the human interaction with the 'battlefield' – then it is possible to argue that the respective strength of China and the USA will depend on the capacity of their politico-economic systems to permit opportunities for entrepreneurial and technically advanced groups, within both the civilian economy and the military, willing to challenge existing practices. The extent to which change is compatible either with authoritarianism or with a capitalist democracy is unclear, and so is the ability of both political systems to create and sustain alliances.

Such alliances are important to military capability. They create commitments and expectations, and also affect the nature and ethos of military organization. This is true of wartime coalitions and of peacetime war preparation, and also, more generally, of the nature of armed services enmeshed in alliances: alliance commands, procurement policies and doctrinal requirements all shape organization. Armed forces within alliances are also, albeit more intangibly, affected by the organizational cultures of other components of such alliances.

Control by and of the military are central to issues of state authority, accountability, and popular participation. Consideration of the possibilities of coups directs attention to the problem of control posed by the military, especially if it does not see itself as apolitical. In many states, for example Indonesia and Pakistan, the issue of control over the military is, and has been, a central feature of politics.[29] It is, however, difficult to integrate into a typology of military organization because circumstances vary so greatly. The political and administrative position of an armed force in a particular society depends heavily on the relative strength and the relative credibility of the other public institutions, and of alternative sources of military strength, most obviously the police. Thus, in Pakistan and Turkey, there has been pressure for the military to take power when civilian politics are discredited.

The size and purpose of the military are set by political factors, and subject to public debate. Such debates have a long pedigree, reflecting a reluctance to leave operational and organizational issues to government. The *Craftsman*, the leading London opposition newspaper, declared on 13 July 1728: 'History does not afford us one instance of a people who have long continued free under the dominion of the sword . . . the very situation of our country, which is surrounded with rocks and seas, seems to point out to us our natural strength, and cut off all pretence for a numerous, standing, land force.'

Control is the concluding theme of this chapter. First, it is likely that governments will seek to retain direct control over the military, and that its organization will remain part of the state. In a period when privatization, or the intermediate stage of 'hiving off' into autonomous corporations, has affected government across most of the world, the military has remained largely immune, at least as far as the fighting 'edge' (as opposed to the logistical backup) is concerned.

If 'privatization' is a theme in the discussion of the modern military, then it must possibly be reconceptualized in order to take note of the move away from conscription, and also of the contractualism on which voluntary service rests. Nevertheless, there are no obvious parallels in modern states to the entrepreneurship of early-modern warfare, the military markets of Europe and India, although there are parallels to one aspect: namely, the subsidization of the armed forces of allies. Modern states might contract out aspects of military training, but they retain control over the process (and pay for it), and the extent to which officership is a function of state-directed training is crucial to modern military organization. Modern states do not use a military market for recruiting officers or sailors: mercenaries generally play a role only when states are weak and subject to civil conflict.[30] There is no ready availability outside the state militaries of the skills and training necessary for modern advanced aerial, naval and land combat operations.

'Privatization', nevertheless, could be discussed in terms of the division of responsibilities and benefits within alliances, for example in the Gulf War of 1990–1. Then, Japan and Saudi Arabia, in effect, paid the USA for the use of its specialized military, in a modern example of protection costing. This organizational division was arguably the most effective aspect of the conflict.

Secondly, any theory of military organization must take note of problems of internal and external control, for organization is not an abstraction: the armed forces are too important in most societies to leave out of the equations of politics, even if only in the wider definition of the latter. In the West, external control became less of a problem over the last half millennium. Instead, the military has become an instrument of state, most obviously in the USA.[31] There, the most powerful military in world history has never staged a coup, and has had relatively little influence on the structure, contents or personnel of politics. A cult of professionalism is central to the ethos of the American officer corps, and their training is lengthy. Washington, Grant and Eisenhower were *elected* presidents, as were other former generals: Jackson, William Henry Harrison, Taylor, Garfield, Hayes and Benjamin Harrison. The last three, like Grant, had been Union generals in the Civil War, while William McKinley had been a Union soldier. The more unstable and self-obsessed Douglas MacArthur was kept away from power.

Yet, civil–military relations have not always been easy, in part because of military misgivings about civilian leadership; and a politicization of the senior military command has been held responsible for poor leadership during the Vietnam period. Furthermore, under President Reagan in the 1980s, there was a development of covert operations outside formal military structures. Still, this is a world away from the political activism and conspiratorial role of, for example, the Black Hand within the pre-First World War Serbian Army.[32] In most of the world there is no equivalent to the requirement, under the Goldwater-Nichols Department of Defense Reorganization Act of 1986, on the Chairman of the Joint Chiefs of Staff to report to Congress on the effectiveness of the armed forces every three years. Furthermore, in 1993 Congress established an independent commission to report on military activity.

The American model of a professionalized force under democratic control has been influential in Western Europe since 1945, in part due to the reorganization of society (and the military) after the Second World War, especially in defeated Germany and Italy, and in part thanks to the influence of the American model through NATO and thanks to American hegemony. The depoliticization of the military has been accentuated with the shift, or return, to professional-based forces over the last three decades. This owes much to military reasons, because conscripts are of limited direct use in high-tech war and, as the Israelis have learned in occupied territories, in constabulary missions as well. Operationally functional, the move away from mass conscript forces also removes the domestic visibility involved in committing reservists in limited wars.

There are wider consequences in terms of organization and politics. Professionalism and voluntary service create a different organizational culture. In addition, professional armed forces can have distinctive politics. This can be seen with the evolution of the American military, once professional, into a force of Republican voters. In the wider political sense, the professional military in the modern West is a crucial tool of governments, thereby enabled to mount or sustain foreign policies their citizens might not support if their sons were doing the fighting.

The situation is very different, however, in Latin America, Africa, the Middle East and South Asia, despite American efforts in states such as Indonesia to encourage the military to accept civilian control. Western-style military establishments have a disproportionate independence and impact in post-colonial systems where nothing else seems to work very well and too many countervailing institutions have lost credibility and authority.

Sometimes, these establishments are the major force holding the post-colonial state together.[33] Nigeria in the 1960s, especially during the civil war of 1967–70, exemplifies this process. In Pakistan, force was employed to maintain control over East Pakistan, particularly in the spring of 1971, but that December, an Indian invasion led to the collapse of Pakistani power and the creation of the new state of Bangladesh. In Burma, a series of challenges led to a serious civil war in 1948–55. In 1959, the Moroccan army crushed a revolt by the Berbers in the Rif mountains. In China, the Tibetan rebellion of 1959 was defeated in 1962 and military rule was imposed, while Muslim revolts in Xinjiang were suppressed in the 1980s. In Syria, where the Ba'th party seized power in a military coup in 1963, the military was used to suppress revolts in the late 1970s and the early 1980s. In Iraq, force was used in the 1990s to suppress both the Kurds and the Marsh Arabs, Southern secessionism in the Yemen was crushed in 1994, and in Libya it is alleged that government forces suppressed Islamic guerrillas in Jebel Akhdar (Cyrenaica) in 1996. In 1998, the Tajik army suppressed a rebellion in the Khojand region of Tajikistan where many Uzbek-speakers live: the Uzbeks backed the rebellion. In some countries, the situation is accentuated by sustained resistance on the part of regions/regional peoples with their own forces, such as the Kurds, or the Karens in Burma.

However, the role of the military in such states leads to a situation in which military professionalism is compromised while society is militarized.[34] The two processes culminated in the frequency of coups. For example, in Africa coups after the Second World War gathered pace with decolonization. Within forty-five years, there were coups in Egypt (1952 and 1954), Sudan (1958, 1985, 1989), Togo (1963), Zanzibar (1964), Zaire (1965), Nigeria (1966), Ghana

(1966, 1972, 1978, 1979), Sierra Leone (1967 and 1997), Mali (1968), Libya (1969), Uganda (1971, 1980, 1985, 1986), Madagascar (1972), Ethiopia (1974), the Central African Republic (1979 and 1981), Equatorial Guinea (1979), Liberia (1980), Upper Volta (1982), Nigeria (1985), Lesotho (1986) and Ivory Coast (1999). In addition, there were a whole series of attempted coups including, for example, in Gabon (1964), Gambia (1981), Uganda (1982) and Nigeria (1990).

Partly as a consequence of the potential political role of the military, their internal structure in much of the world has often been a matter of the creation and maintenance of competing, even adversarial, structures in order to limit the possibility of coups. Thus, in Saudi Arabia there has been a counterpointing of National Guard and the army, in Egypt of Central Security Forces and the army, in Syria of Defence Regiments and the army, in South Yemen of militia and the army, in Iraq of Republican Guard and the army, and in Russia of Ministry of Internal Affairs troops and the army. In 1962, Soviet troops refused to fire on rioters at Novocherkassk who had been enraged by food shortages and wage cuts. Eventually, Ministry of the Interior troops restored control.[35]

The creation of militarized structures that are not themselves the formal military creates specific problems for the latter, and can also ensure that force is wielded by a range of bodies, some of which slide off into the realms of political violence and/or organized crime. This process is far less common in specialized weaponry services: naval, air and missile units cannot be readily employed in internal policing. It is, however, also used in the 'intelligence' field, as in Syria.

Dictatorial systems have been particularly prone to seek to use, but also neutralize, the military. In Nazi Germany, the SS was expanded to include many military units, in order to provide a Nazi counterpoint to the army. The SS itself was a development of the notion of the 'political soldier' devised by the Freikorps after the First World War. The Freikorps was the model, but the SS was bureaucratic and under state control. In Taiwan, Chiang Kaishek made his son Jingguo, who already controlled the secret police, Minister of Defence in 1965 in order to ensure family power.

This politicization of military and pseudo-military structures does not pertain in the modern West, or, if so, to a smaller degree. There are functional differences in the West between units and branches that are linked to questions of strategic doctrine, and thus to the 'political' grounds of military purpose, but these questions and differences do not focus on the internal stability of the state. This definition of the West, however, does not include Latin America.

In many parts of the world, the ability of the state to monopolize force has broken down. Regional resistance and the limited legitimacy of national regimes can combine in a very damaging fashion. State-organized war is a notion that, for example, means little in modern Sierra Leone and Liberia. In both, organized armed forces under central control have been largely replaced by drugged teenagers and outright looters who have little, if any, idea of the cause they are fighting for, except for their personal gains. Political objectives beyond the capture of power are hazy, and 'wars' are financed primarily by criminal operations and forced extortions. There are no chains of command or (often) even uniforms that distinguish 'troops' from each other or from other fighters.[36] In other states, there are powerful militias, for example the Hutu *interahamwe* in Rwanda, that challenge all control.

The organizational model that best helps analyse such armies is that of the criminal gang, a modern echo of St Augustine's comparison of Alexander the Great to a brigand: 'in the absence of justice there is no difference between Alexander's empire and a band [*societas*] of thieves'.[37] Frederick Lane's theory of protection costs[38] takes on a different meaning in such contexts. Ironically, Frederick the Great compared himself in 1757 to a traveller attacked by a gang of brigands.[39] In response, the West today deploys unofficial or independent forces, such as Sandline International, a British mercenary force active in Africa and elsewhere in the late 1990s, or employs regular forces in what are termed peacekeeping missions.

As an extension of the situation in countries such as Sierra Leone, it is unclear how best to treat the modern forces of sub-sovereign bodies, both those with an established international position and territorial base, such as the Palestinian Authority, and those that lack either, such as the IRA. It is not always clear how far criminal movements, such as the Mafia in Sicily or drug cartels in Colombia, can be distinguished from insurrectionary forces. The guerrilla FARC movement in Colombia relies on drug money. Thus, there is frequently no clear division between lawlessness and rebelliousness. Furthermore, the response on the part of government forces is not always easy to define. In some countries, in addition, the forces of order have been and are linked to semi-official or unofficial vigilante groups. Thus, Chiang Kaishek used the Green Gang to support the army against the Communists in Shanghai in 1927, and gave its leaders high military ranks,[40] while in Venezuela in 1999, alleged criminals were attacked by residents, with the apparent connivance of the local police in many cases. This was not, of course, civil war, but an aspect of the difficulty of defining 'sides'.

While in Burma and Peru insurrectionary movements derive much of their revenue from criminal activities, in some states, such as China, Indonesia and Nigeria, sections of the armed forces intervene actively in the economy and are widely suspected of corrupt practices. This is an aspect of their active political role that is far from new. In the late 1930s, the Japanese army was closely linked to the Manchuria Heavy Industry Company, and this underpinned the Japanese military commitment in China. In the case of China, among other states, it is possible that military commanders will gain regional power, not least because units are not rotated. This may lead to the recurrence of the warlord phenomenon seen in 1916–36, and, indeed, on earlier occasions, for example the 1660s–1670s.[41]

Contrasts in organizational structure and ethos, and in the political position of the military, are unlikely to alter in the near future. The role of the military as the focus of power (and of social mobility, and even the expropriation of wealth) will remain in many states. It is difficult to believe that the ethos of their organization will be primarily set by external military purposes.[42] This will be even more the case if the hegemonic power of the USA can maintain international peace, as in Latin America. The military will therefore acquire advanced weaponry, as if preparing for foreign war, but will primarily fix its sights on domestic goals, in order to safeguard its essentially authoritarian view of the fatherland.

These goals have prevailed for many years in Latin America. For example, there were numerous military coups, including in Mexico in 1920, Chile in 1924, 1925, 1932 and 1973, Argentina in 1928, 1943, 1955, 1962, 1966, 1970 and 1976, the Dominican Republic in 1930 and 1963, Guatemala in 1931, 1944, 1954, 1978, 1982 and 1983, Paraguay in 1936, 1937 and 1989, Peru in 1948, 1962 and 1967, Venezuela in 1948 and 1958, Bolivia in 1951, Colombia in 1953 and 1957, Brazil in 1964, Grenada in 1983, and Ecuador in 2000. In addition, there were military rebellions, for example in Brazil in 1922, 1924, 1932 and 1954, Peru in 1930 and Argentina in 1988. Military force was also used against dissidents, for example peasant opposition in Mexico in 1929, El Salvador in 1932 and Honduras in 1932, 1937 and 1944, an insurrection in Cordoba in Argentina in 1969, and widespread resistance in Nicaragua and El Salvador in the 1970s and 1980s. In 1999, para-military Chilean Special Forces police were deployed against Mapuche Indians who were in dispute with the powerful forestry industry.

This list could readily be extended to demonstrate further the frequent use of force to maintain control and wield power. Armies have served to intimidate the population. This was particularly true of military regimes, for example Chile after the 1973 coup, but also of civilian governments, such as those of Peru and Colombia in the 1980s and 1990s. In 1992, the Peruvian President used the army to shut down the country's Congress and courts. The military was also used for policing. In Jamaica, for example, the army was employed in 1999 on joint patrols with the police.

In Argentina, the sole South American state to fight a major foreign war in the last half-century, the military's role in the country's history was essentially that of a domestic political force, rather than a contestant with Britain for the Falklands/Malvinas in 1982 or a shadow-boxer with Chile over frontier differences.[43] The deficiencies of the Argentinian army in 1982 reflected the absence of combat-readiness on the part of poorly trained conscripts and their officers, and the lack of an understanding of the operational requirements of modern infantry conflict, specifically small-unit cohesion.

Armies in such circumstances can be regarded in part as spoils systems, in which control over force is employed to the profit – political, social and financial – of their members, especially the leaders. This was also the situation in early-modern Europe, but it is not so with the modern Western military. The leaders of the latter enjoy handsome and reliable salaries, but scarcely the economic and political power of, for example, their Indonesian and Chinese counterparts, to select two states that are not under military rule. In such systems, promotion can owe little to professional criteria, and status is a matter of politics and ethnicity as much as training and competence. Furthermore, an emphasis on internal control and order helps ensure that the army receives more resources from the air force and navy in states such as Indonesia and Argentina.

Thus, to reverse the usual account, it is necessary to look for narratives and paradigms of military development primarily in terms of contrasting political cultures, rather than technological 'progress'. Any account of military organization that focuses principally on the effectiveness of weaponry will be limited. Instead, in order to understand the nature and development of such organizations, it is crucial to examine political contexts, configurations and purposes.[44] These political needs and constraints reflect different societies and cultures, and the latter hinder any understanding of military development in terms of conformity to, or divergence from, a uniform global model, whether set by technology or by other operational factors.

NOTES

1. PRO WO 33/1819, p. 50.
2. S. Lone, *Japan's First Modern War: Army and Society in the Conflict with China 1894–1895* (1994).
3. J.W. Geary, *We Need Men: The Union Draft in the Civil War* (De Kalb, IL, 1991).
4. M.S. Coetzee, *The German Army League: Popular Nationalism in Wilhelmine Germany* (Oxford, 1990).

5. W.A. DePalo, *The Mexican National Army, 1822–1852* (College Station, TX, 1997).

6. Foreign Manoeuvres, 1898, PRO WO 33/2819, p. 26.

7. Foreign Manoeuvres, 1896, PRO WO 33/2816, p. 43.

8. Foreign Manoeuvres, 1900, PRO WO 33/2822, p. 93.

9. Foreign Manoeuvres, 1900, PRO WO 33/2822, p. 10.

10. J.F. Godfrey, *Capitalism at War: Industrial Policy and Bureaucracy in France 1914–1918* (Leamington Spa, 1987); J. Horne (ed.), *State, Society and Mobilization in Europe During the First World War* (Cambridge, 1997).

11. J.W. Chambers, *To Raise an Army: The Draft Comes to Modern America* (New York, 1987).

12. Jenkinson to Amherst, 24 October 1780, PRO WO 34/127 fol. 155.

13. PRO WO 33/1512, p. 3.

14. G.Q. Flynn, *The Draft, 1940–1973* (Lawrence, KS, 1993).

15. P.W. Schroeder, *The Transformation of European Politics 1763–1848* (Oxford, 1994), p. 125.

16. B. Gökay, *A Clash of Empires: Turkey Between Russian Bolshevism and British Imperialism, 1918–1923* (1997); S. Cronin, *The Army and the Creation of the Pahlavi State in Iran, 1910–1926* (1996); R. Potash, *The Army and Politics in Argentina, 1928–1945* (Stanford, CA, 1960); R.M. Levine, *The Vargas Regime: The Critical Years, 1934–1938* (New York, 1970).

17. H.W. Nelsen, *The Chinese Military System: An Organizational Study of the Chinese People's Liberation Army* (Boulder, CO, 1977); E. Joffe, *The Chinese Army After Mao* (Cambridge, MA, 1987).

18. R. Wells, *Insurrection: The British Experience, 1795–1803* (Gloucester, 1983), pp. 79–109; A.K. Wildman, *The End of the Russian Imperial Army: The Old Army and the Soldiers' Revolt, March–April 1917* (Princeton, NJ, 1980); L.V. Smith, *Between Mutiny and Obedience: The Case of the French Fifth Infantry Division During World War One* (Princeton, NJ 1994).

19. BL Add. vol. 13579 fols. 7–8.

20. R. Murphey, *Ottoman Warfare 1500–1700* (1999).

21. R.E. Ford, 1968. *Understanding the Surprise* (1995); J. Prados, *The Hidden History of the Vietnam War* (Chicago, IL, 1995); R. Brown, 'Limited War', in C. McInnes and G.D. Sheffield (eds), *Warfare in the Twentieth Century: Theory and Practice* (1988), pp. 177–84.

22. S.P. Rosen, *Winning the Next War: Innovation and the Modern Military* (Ithaca, NY, 1991), pp. 35–61.

23. D.B. Ralston, *Importing the European Army: The Introduction of European Military Techniques and Institutions into the Extra-European World, 1600–1914* (Chicago, IL, 1990); K. Fahmy, *All the Pasha's Men: Mehmed Ali, His Army and the Making of Modern Egypt* (Cambridge, 1997); J. Dunn, 'Egypt's Nineteenth-century Armaments Industry', *JMH*, 61 (1997).

24. S.N. Sen, *Anglo–Maratha Relations 1785–96* (Bombay, 1994), pp. 294–5.
25. This is most accessible to English readers through a translation by Michael Howard and Peter Paret (Princeton, NJ, 1976).
26. C. Bassford, 'Wellington on Clausewitz', *Consortium on Revolutionary Europe: Proceedings, 1992*, p. 389.
27. C. Schmitter (ed.), *Military Rule in Latin America: Function, Consequences and Perspectives* (1973).
28. D.G. Muller, *China as a Maritime Power* (Boulder, CO, 1983); J.W. Lewis and X. Litai, *China Builds the Bomb* (Stanford, CA, 1988) and *China's Strategic Seapower: The Politics of Force Modernization in the Nuclear Age* (Stanford, CA, 1995).
29. S.P. Cohen, *The Pakistan Army* (Berkeley, CA, 1984); B. Cloughley, *A History of the Pakistan Army: Wars and Insurrections* (Oxford, 1999); H. Crouch, *The Army and Politics in Indonesia* (2nd edn, Ithaca, NY, 1988); A.H. Young and D.E. Phillips (eds), *Militarization in the Non-Hispanic Caribbean* (Boulder, CO, 1986); B.W. Farcau, *The Transition to Democracy in Latin America: The Role of the Military* (Westport, CT, 1996).
30. G. Teitler, *The Genesis of the Modern Officer Corps* (1977).
31. S.P. Huntington, *The Soldier and the State: The Theory and Practice of Civil–Military Relations* (Cambridge, MA, 1957); C. Welch, *Civilian Control of the Military* (New York, 1976). For an example of a very different tradition, see G. Craig, *The Politics of the Prussian Army 1640–1945* (Oxford, 1955).
32. R.F. Weigley, 'The American Military and the Principle of Civilian Control from McClellan to Powell', *JMH*, 57 (1993), and 'The Soldier, the Statesman and the Military Historian', *JMH*, 63 (1999); R.H. Kohn, 'Out of Control: The Crisis in Civil–Military Relations', *National Interest*, 35 (1994); A.J. Bacevich, 'The Paradox of Professionalism: Eisenhower, Ridgway and the Challenge to Civilian Control, 1953–1955', *JMH*, 61 (1997).
33. S.E. Finer, *The Man on Horseback: The Role of the Military in Politics* (1962).
34. J.S. Ikpuk, *Militarisation of Politics and Neo-Colonialism: The Nigerian Experience* (1995); J. Peters, *The Nigerian Military and the State* (1997).
35. R.R. Reese, *The Soviet Military Experience: A History of the Soviet Army, 1917–1991* (2000), p. 141.
36. A. Clayton, *Factions, Foreigners and Fantasies: The Civil War in Liberia* (Sandhurst, 1995).
37. Augustine, *City of God*, IV, 4.
38. F.C. Lane, *Profits from Power: Readings in Protection Rent and Violence-Controlled Enterprise* (Albany, NY, 1979).
39. Frederick to Margravine of Bayreuth, 22 July 1757, *Politische Correspondenz Friedrichs des Grossen* (46 vols, Berlin, 1879–1939), vol. 15, p. 261.

40. B.G. Martin, *The Shanghai Green Gang: Politics and Organized Crime, 1919–1937* (Berkeley, CA, 1996).

41. D.S. Sutton, *Provincial Militarism and the Chinese Republic: The Yunnan Army, 1905–25* (Ann Arbor, MI, 1980); E.A. McCord, *The Power of the Gun: The Emergence of Modern Chinese Warlordism* (Berkeley, CA, 1993).

42. For an example of the role of domestic factors, specifically the search for a prestigious image to dissuade opposition, in encouraging the use of non-utilitarian uniforms and other practices, see S.H. Myerly, *British Military Spectacle from the Napoleonic Wars through the Crimea* (Cambridge, MA, 1996).

43. B. Loveman, *For 'la Patria': Politics and the Armed Forces in Latin America* (Wilmington, DE, 1999); D.L. Norden, *Military Rebellion in Argentina: Between Coups and Consolidation* (1996).

44. C.C. Demchak, *Military Organizations. Complex Machines* (Ithaca, NY, 1991).

FOUR

The Question of Technology: The Early-Modern Example

The issue of technology in conflict is one that interests most commentators on war. It plays a major role in accounts of military change. This is especially true of developments in weaponry, although they are only a part of the question of the role of technology. Arguably, these developments attract far too much attention, in discussion both of technology and of war. Discussion of the role of technology commonly proceeds from particular military contexts or epochs, but more generally, it can be seen as an aspect of the problem of assessing changes in material culture. This is as true of the long period of time delimited in terms of such culture – the Bronze Age, Iron Age and so on – as of more recent centuries. In the former case, the difficulty of recovering the culture of war when only the archaeological record is present ensures that material aspects are heavily stressed, and the material remains determine assumptions about social and political structures, and about the purposes and character of conflict.

In order to assess military technology, it is best to turn to periods for which a variety of types of source survive. This book now turns to an account of the major issues and developments in the early-modern period, and continues by looking in greater detail at European expansion. These developments are significant for the situation today as they are seen as the onset, and in part cause, of modernity, not least in military matters.

90

The variety in the number of possible approaches to the question of the role of technology in early-modern warfare not only poses a problem, but also ensures that a methodology is offered by whatever prioritization is selected, an observation that is also pertinent for the present day. Should we consider this question first, for example, from the perspective of early-modern warfare, or from that of early-modern technology? Alternatively, should our priority be that of military history as a whole? If so, are we going to put the emphasis on the early-modern period, however understood, as an age of change? In short, if we move the focus from the early-modern period to that of warfare as a whole, does our analytical focus and meta-narrative necessarily concentrate on change? The same question may arise if we turn to technology.

First, it is necessary to ask what is meant by 'early-modern', and how far our understanding of the term leads to an attempt to define a separate period that is then given special characteristics and reified in order to endow it with causative powers. This is a general problem that relates far more widely than simply to military history, but that also greatly affects our understanding of the latter. For this reason, this and similar problems need to be addressed in works on military history. Consideration of the concept of the early-modern is one approach to the issues of modernity and modernization, both of which are important for military (and other) history, and each of which greatly affects assessment of present developments and future trends.

Development of the notion of the early-modern essentially reflects two intellectual thrusts: first, Eurocentricity, and second, the Cronos problem, more specifically the tendency in European intellectual culture from the Renaissance to abandon the practice of viewing past 'golden ages', and instead, to disparage preceding ages, or at least, regard the present as better than the past, and a graspable future as likely to be better than both. This tendency focused in Europe on the grand tradition of bashing *the* Middle Ages. The latter were defined in order to be castigated and to serve as a counterpoint to what came after. The Renaissance was, in part, also understood as a recovery of the Classical period, but this was sufficiently distant and obscure not to serve as a real model for a tendency that was not in fact conservative.

In short, with the concept of the early-modern, we are in the world of primitivization and foundation myths, more particularly the foundation myth of the modern world: in military terms, gunpowder with a bang. For a long time, the modern world was seen as depending on a rejection of what had come before, rather than representing a development of, or from, it. The force of this rejection created the turning point.

This intellectual thesis was important to the public cultures of the Europeanized Western world. Christianity was more central to intellectual life than we today appreciate, and in Protestant Europe, the early sixteenth century was seen as the turning point. There was something to reject, and a clear moment of rejection. Furthermore, the moment had meaning for Christian intellectuals, whether religious or secular, even if their interpretation of it was different. Whether they looked to the Renaissance or to the Reformation, or both, these developments, as generally understood, were taken to mark a dramatic break.

It was not surprising that this break was then read into other contexts: economic, social, political and cultural. Furthermore, it took on meaning subsequently in terms of the structure of the academic profession in the West. In a number of subjects, not only history, but also, for example, literature, theology and philosophy, clear chronological divisions were drawn and then entrenched in professional careers, publication strategies and teaching syllabuses.

If, however, this perspective on the early-modern period is questioned, then it becomes possible both to disaggregate its components, and to scrutinize particular developments without employing an explanatory framework that relies, in part, on a possibly misleading general theory of the early-modern period. Specifically, it is possible to query the notion of modernization and also the idea that changes, at least in part, are to be explained by a process of emulation and of diffusion from other spheres. So, our positioning of military history, military change and the 'Military Revolution', all three of which are very different, depends in part upon our account of the early-modern period, specifically upon notions of post-medievalism, of modernity, and of modernization.

At one level, the narrative of military revolution is the counterpart of humanism, printing, European exploration, the Renaissance, the Reformation, and a new mental world. As such, it is challenged by arguments that the onset of modernity and of modernization should be dated far later, either to the second half of the eighteenth century, with its cult of secular progress, or to the nineteenth, with the astonishing transformation in the productive and organizational capability of the Western world. The former change is often referred to as the Enlightenment, and the latter as the Industrial Revolution, although neither is a precise fit.

This issue is also of particular importance for the discussion of technology. It has been argued, for example by Leo Marx, that the term 'technology' is inappropriate before the nineteenth century. Instead, from the sixteenth to the eighteenth century, there were significant changes in the invention and use of machines in the West (and this would extend to the military sphere of gunpowder), but that 'technology', in the sense of the full integration of the invention, development and use of machines into society and the economy, was conditional upon a prior social revolution in which earlier social divisions had been blurred, rapid social ascent was possible, financial capital was readily available, and mobility of labour was established. At that point, Marx argues, 'technology' can be discerned; in other words, it is to be seen as part of an entire socio-economic system.[1]

Furthermore, it is possible to re-examine the role of force in the narrative of the early-modern period. The notion of that period, and of the transition from medievalism, depends both on an understanding of force as playing a crucial role, and on a presentation of the nature and use of force as novel. Thus the idea of a Military Revolution takes a part akin to that of the transition from feudalism to capitalism in Marxist and Marxisant models. This helps account for the popularity of the idea of such a revolution, and also for the need for a theory to explain it. If the concept did not exist, it would be necessary to devise an alternative to explain

93

major shifts that have been discerned in the world structure of power, in state-formation within Europe, and in some accounts elsewhere, and in the world economy. War can be fitted into modernization theory. More generally, force apparently both explains and is the *modus operandi* of what has been seen as a politics and culture of control and expropriation.[2] Changes in the latter thus have to be linked to military developments. The Military Revolution, therefore, apparently fulfils the need for an abrupt and violent close to medievalism, the Middle Ages, and for the beginning of a world in which Europeans were on a path towards global influence, if not dominance.

From Michael Roberts onwards, proponents of the thesis of the Military Revolution have argued that certain developments in the conduct of war (*partly* driven by technology, but far from solely so) had very large consequences, both for the European state structure and on the course of the evolution of later Western military structures and systems. This does not require an abrupt close to the Middle Ages and, for Roberts, 'the' Military Revolution began in 1560, after the Middle Ages are generally seen as over, but, as employed in much of the literature, gunpowder, like printing, became a totem of a different world.

As already suggested, there are conceptual problems with this thesis, aside from the military dimension. It makes only very limited sense in terms of much revisionist work on sixteenth- and seventeenth-century European history. Against the notion of new monarchies employing cannon and professional forces to monopolize power, this work emphasizes the continued strength of the nobility.[3] Thus, without as yet considering war, it is becoming clear that, at least in terms of the revisionist views already outlined, the military history of the period that has to be explained is very different, in terms of cause, course and consequence, to that outlined by Roberts and required by his readers.

There is also the global dimension. This is no mere footnote to developments within Europe. The early-modern period at the world scale can be dovetailed into the modernizing view of the period in Europe if emphasis is placed upon European expansion, and/or if the situation outside the European world is treated as presenting a parallel or, rather, series of parallels. If the latter approach is adopted, it carries with it the apparent additional merit that the Europeans thus succeeded because they were better at following the same course that the other major powers were on. The reasons for European military success are thus employed to establish a global hierarchy of military proficiency, encouraging a neglect of the non-European powers, or a treatment of them as failures and/or victims.

These approaches, however, can be questioned. Historians who argue for a technologically driven Military Revolution as the key causative factor behind the European rise to dominance in the early-modern period are mistaken, because, first, although the Europeans created the first global empires, this was only partly due to the military developments of the Military Revolution and, second, there was no European dominance. Third, the crucial changes in European armies in this period can be discussed in organizational as much as technological terms.

The revolutionary character of European overseas military activity can be questioned. Discussion of this activity commonly focuses on the Portuguese in the Indian Ocean and the *conquistadores* in the New World. These, however, should not be regarded as more generally indicative, both for methodological and for empirical reasons. The general concept of the Military Revolution encourages a uniform perception of the world as a single global (systemic) entity in which there are variations in the extent of European success, but the central issue is the character and use of European power. However, although the notion of a single system is a tenable perception for Europe, from the period of the use of the notion of the balance of power, it encounters the difficulty that, until the eighteenth century, systemic perception of the world by Europeans appears to have distinguished sharply between the Old World and the New. Hence, European warfare in much (but not all) of Africa and Asia was private and conducted under the supervision of chartered trading companies. As a consequence, the use of armed force was much more instrumental, and governed by issues of cost-effectiveness, than in spheres where direct governmental control was the case.

It is also unclear that it is appropriate to read across from one success (or a group of successes) in order to suggest a shift in global military capability. Furthermore, it is probably mistaken to consider naval alongside land warfare. The trajectory, causation and limitations of European naval success[4] were different to the case on land, although they were not unconnected. In addition, there was a more direct linkage, in that enhanced naval capability permitted a far greater range of land operations. In the period from *c*. 900, the Europeans had been able to apply force across the Mediterranean, but from the late fifteenth century, it was a case of force applied across the oceans. Modernity has been seen as arriving in the shape of a heavily armed warship capable of sailing long distances. It is important, however, not to assume that the development of rigging, muzzle-loading cast-metal guns, and heavy guns on the broadside, all occurred at once, and were mutually dependent. Indeed, by the mid-fifteenth century, galleys were being built to carry guns, and in 1513 these showed their ship-killing capability at the expense of the English fleet off Brest.

Naval range created the possibility of distant land operations, but could not ensure their success. European failures on land qualify any account emphasizing a capability gap at the expense of non-Western powers; but it is anyway unclear whether an aggregate measure of achievement is most pertinent. Such a measure fails to address the multifaceted nature of European expansion and military activity, and the different nature of tasks and expectations in particular areas. The same point could be made for other systems operating in a number of different contexts, such as the Ottoman Turks, or the Mughals in the seventeenth century and the Manchus in 1660–1800.

What then of the attempt to suggest parallels, most famously in the use of gunpowder?[5] Thus, for example, it is possible to incorporate the Ottoman overrunning of Syria, Palestine and Egypt in 1516–17 or the Moroccan conquest of Timbuktu in 1591 into a model that otherwise focuses on the *conquistadores*. This can also be extended to South and East Asia, and employed to discuss the internal as well as the external use of force.

There are, however, problems with this analysis. It assumes that firepower, and more particularly gunpowder firepower, was crucial to the military successes of the powers discussed. This is unclear, in part because of the lack of empirical research, or its contentious nature. This is why advocates of the early-modern Military Revolution work with the big picture and overall trends, employing individual cases as illustrations more than evidence. However, this method invites questions about the applicability and contextualization of the illustrations cited, and thus directs attention to the issue of how overall trends are to be judged. These methodological points are more generally true, both of military history and of the military situation today, for example of the role of air power.

It is anyway unclear why, if the earlier introduction of gunpowder (as of printing) in China did not lead to revolutionary changes, the situation should have been different, or be regarded as different in Europe. Gunpowder provided the basis for different forms of hand-held projectile weaponry and artillery on land and sea, but the technique of massed projectile weaponry was not new. More specifically, it is possible to regard gunpowder weaponry as an agent, not a cause, of changes in warfare.

Related to this is the need to consider how best to incorporate into any description of military systems or analysis of military change armies that do not correspond to the meta-narrative, whether technologically driven or not. The armies in question were frequently not infantry-based, and some of them cannot be regarded as the forces of settled societies and/or 'advanced' bureaucratic states. This is true, for example, of the Safavids and Mughals at the outset of the sixteenth century, of the Uzbeks later in the century, the Marathas in India in the seventeenth and first half of the eighteenth century, and the Afghans in the eighteenth century. Those African armies that placed only limited emphasis on the use of gunpowder can also be discussed under the same heading.

Any consideration of such forces invites a re-evaluation of comparative global military advantage and capacity, of the extent to which it is appropriate to consider military progress in terms of the use of gunpowder, particularly siege artillery and, subsequently, the use of gunpowder by infantry employing volley fire. It is also necessary to reopen the question of how best to measure success, specifically the extent to which conquest and the territorialization of power are the crucial criteria. Instead, it is possible, for example, to place weight on successful raiding. The issue of how best to measure success relates to that of military effectiveness, through the question of how far capability should be considered in terms of the fitness for purpose of particular forces. This fitness was, and is, constructed 'culturally' in socio-political terms, rather than being technologically driven.

A rebuttal of a simple theory of gunpowder triumphalism does not have to extend to a denial of any role for such weaponry, but it does invite a searching enquiry as to their antecedent role in the process of change. This also leads to the question of what technology means in the context of this question. In particular, there is the issue of how far and how best to distinguish between changes in weaponry and their wider context, specifically the training, systematization and industrial production that stemmed from the large-scale deployment of such weaponry on land and sea. This is all linked to technology. In addition, if we see the major technological innovations in Europe as occurring prior to 1500 and argue that in the sixteenth century there was essentially incremental change, frequently at a sluggish rate, but an increase in the number of gunpowder weapons, then we have to decide how to rank the relative importance of these two processes. The latter made gunpowder weaponry normative, which is to say that drill, tactics and assumptions were all focused in a particular way. Yet elsewhere in the world, the focus was different, the standardization less pronounced. This can be treated as failure, but it is a little difficult to regard the Manchus, for example, in this light. Indeed, tactical flexibility in some contexts meant a mix of weaponry very different from that of standard European armies.

The question of how best to prioritize different but related developments can also be seen by considering both the role of counter-measures and the impact of enhanced non-weaponry capability. The first lessened the affect of specific weapons and of particular tactics designed in part to maximize their role. Thus, for example, in the eighteenth century the Duke of Marlborough (the leading British general in the War of the Spanish Succession, 1702–13) and later Frederick the Great of Prussia were masters of the delivery of well-timed, concentrated force on the battlefield, but their opponents learned how to predict their plans and to respond.

Nevertheless, thanks in part to enhanced non-weaponry capabilities, the conflicts of the eighteenth century were frequently decisive. The Austro-Sardinian victory at Turin (1706) drove the French from Italy, French victories at Almanza (1707) and Brihuega (1710) won Spain for the Bourbons, and Peter the Great's victory over the Swedes at Poltava (1709) was followed by the Russian conquest of Livonia. The armies (and navies) were better supported and more effective than those of the classic period of the Military Revolution, and the governments were better able to sustain wars in which reasonably well-supplied forces could be directed to obtain particular goals, rather than to have to search for food.

The enhancement of capability within a given technology is an important aspect of military development. For example, the British navy developed its capability in 1714–39, with the creation of bases at English Harbour and Port Antonio in the West Indies, and Mahon and Gibraltar in the Western Mediterranean. This increased the relative margin of operational capability. Furthermore, successive wars in the eighteenth century indicated that the British ability to man its expanded navy (albeit slowly) contrasted with France and Spain over a long conflict, in short that the British navy was more robust than its rivals, even though its technology was similar; indeed, French warships tended to be better built. The expansion of British trade increased the number of sailors available for the navy. As a result of these factors, by mid-century the navy and British maritime strategy were capable of having an impact which they did not possess fifty years earlier.

It is also helpful to have an understanding of tactics and strategy, particularly the latter, in which they are not seen as arising largely from the question of how best to use weaponry. Thus, as already suggested, the question of fitness for purpose has to be addressed, both in terms of specifics – particular armies and states at distinct moments – and more generally. Both, in part, rest on our understanding of the external and internal dynamics of polities. A Darwinian or 'realist' approach that sees continual conflict, or at least confrontation, focuses on the maximization of force; but any assessment that directs greater attention to the role of choice in deciding whether and how to wage war adds a different dimension of complexity. An emphasis on choice necessarily has implications for the discussion of strategy.

In general, the shift in recent years has been away from schematic accounts of state development and international systems, with their inherent determinism, and towards a more fluid understanding of trajectories; although this shift does not comprehend all scholars. Given this context, it is possibly more helpful to focus on the potential of technology and the potentiality of innovation, rather than any crude measure of both, and thus to consider complex questions of adoption, dissemination and adaptation. The latter parallels interest in acculturation as an aspect of imperial conquest and control. It also draws attention to the limitations of devising a typology of warfare simply in terms of symmetrical and asymmetrical confrontation and conflict. The role of potential, and the choices latent in such an approach, lessen the notion of similarity attaching to symmetrical warfare. It also permits us to understand more clearly the varied success of the Europeans in asymmetrical warfare against different opponents. Rather than seeing this largely in terms of the relative difficulty of the task (in terms of opponent and/or ecology), it is possible also to focus on differential ability to utilize the potential of the European military system, as well as varied understandings of what this potential entailed, and the processes of choice involved in understanding and utilization.

As already emphasized, this potential should be seen largely in cultural terms. Adoption of weaponry, and even tactics, can require major changes. These were both within the military – for example, drill and discipline, or meritocratic promotion, or a departure from organization in terms of ethnic difference – and sometimes in society as a whole. The latter was especially the case with the creation of conscript systems, as in Prussia and Russia in the eighteenth century. Yet such changes could be actively resisted, both within the military and in society as a whole, because there was an unwillingness to revise the cultural understanding of the purpose and use of force, military authority, the nature of merit, and the character of command.

These cultural issues are important, but it is important not to overload them as an explanatory device. For example, it can be argued that, for cultural reasons, European forces acquired a tactical organizational superiority in the early-modern period that enabled them to use gunpowder weaponry tactics more effectively than their opponents. Such an argument would focus on the ability to keep cohesion and control in battle, and to make effective use of units. This disciplined unitization of armies (and navies) may be generally discussed in terms of the Renaissance. Developments in ballistics can be linked to the seventeenth-century Scientific Revolution and seen as an aspect of an understanding of cause and effects and a determination to take predictable advantage of this relationship.[6]

This argument is not without value, but it suffers from a Eurocentricity of method. This is twofold. First, European methods of discourse are applauded, specifically the ready reproduction of ideas and instructions through a culture of print, although it is, of course, difficult to assess the impact of this print culture for the conduct of war in the early-modern period. Second, there is a primitivization of the conceptual methods and organizational effectiveness of other forces that essentially rests on our ignorance.

Indeed, we have sufficient sources to show that it was possible for non-European cultures to devise new military solutions. This was true of native warfare in North America. For example, the Red Stick Creeks sheltered behind a formidable log and dirt barricade at Tohopeka in 1814, and this barricade proved resistant to the 3 and 6 lb solid shot of Andrew Jackson's cannon. As a consequence, it proved necessary to storm the position in the battle of Horseshoe Bend. Thus, the destruction of the power of the Creeks cannot be understood as a firepower triumph; indeed, Jackson benefited from the support of White Stick Creeks and Cherokees.[7] There are also enough sources for seventeenth- and eighteenth-century coastal West Africa to show that local states were capable of organizing their armed forces, both comparably to Europeans and in such a way that they could introduce change.[8]

These comparisons relate to coherent states, albeit not always states within a territorialization comparable to the spatial character of European states. It is also necessary to consider the numerous peoples of the early-modern period who lived in less coherent political structures, and to assess what the question of technology meant in their case. Many such peoples practised dispersed styles of warfare, not as a consequence of deep reasoning, but because it was a transition from hunting and a consequence of low population tension. In such a context, the spread of new weapons might have been useful for hunting and warfare, but did not necessarily lead to any social or organizational changes. The availability of firearms, shot and powder, however, became an important issue.

Aside from looking outside Europe, it is also pertinent to look at the military history of the Continent, both in the early-modern period and indeed earlier. This suggests that a focus on technological superiority, and those held to exercise it, may be misplaced. For example, in the pre-firearms era, far from the feudal cavalry and castles of France and Germany making other military traditions, such as Anglo-Saxon England, redundant,[9] it is possible to note strengths in the latter and to explain defeats such as Hastings (1066) other than through weaponry.

This can be repeated for the firearm era, for example by noting the continued impact of Celtic warfare, with its emphasis on shock. More generally, rather than arguing that firearms revolutionized warfare, it is possible to focus on the way in which they slotted into existing tactical systems. The theme then is on adaptation and the combination of arms. This leads to a new chronology: 'The effective tactical synthesis of infantry, cavalry and artillery achieved in the Great Italian Wars of 1494–1529 was contingent on the organization and sophistication of the combined arms approach of fifteenth-century armies, most strikingly the Burgundians.'[10]

Thus, again, the methodological problems of devising a general theory of military capability and change in the early-modern world emerge. These problems can be refocused by adopting a cultural interpretation that considers such 'non-military' factors as reasons for conquest or for the avoidance of aggressive warfare and conquest. Technological and organizational issues can then be seen as enablers, military multipliers that help to explain moments of success, but not the moods that led to the quest for conquest.

NOTES

1. See, for example, M.R. Smith and L. Marx (eds), *Does Technology Drive History? The Dilemma of Technological Determinism* (Cambridge, MA, 1994), and L. Marx, 'In the Driving-seat? The Nagging Ambiguity in Historians' Attitudes to the Rise of "Technology"', *Times Literary Supplement*, 29 August 1997, pp. 3–4.
2. See, for example, J.B. Harley, 'Rereading the Maps of the Columbian Encounter', *Annals of the Association of American Geographers*, 82 (1992), pp. 522–42.
3. Contrast M. Roberts, *The Military Revolution, 1560–1660* (Belfast, 1956) with, for example, W. Beik, *Absolutism and Society in Seventeenth-century France: State Power and Provincial Aristocracy in Languedoc* (Cambridge, 1985), and N. Henshall, *The Myth of Absolutism: Change and Continuity in Early Modern European Monarchy* (Harlow, 1992).
4. For naval development within Europe, see N.A.M. Rodger, 'The Development of Broadside Gunnery 1450–1650', and G. Parker, 'The Dreadnought Revolution of Tudor England', *Mariner's Mirror*, 82 (1996).
5. W.H. McNeill, *The Age of Gunpowder Empires 1450–1800* (Washington, DC, 1989).
6. B. Steele, 'Muskets and Pendulums: Benjamin Robins, Leonhard Euler and the Ballistics Revolution', *Technology and Culture*, 34 (1994), pp. 348–82.
7. J.A. Reid, 'Andrew Jackson's Victory at Horseshoe Bend', *Journal of the War of 1812*, II, 3 (1997), pp. 2–3.
8. J.K. Thornton, *Warfare in Atlantic Africa, 1500–1800* (1999).
9. R. Bartlett, *The Making of Europe: Conquest, Colonization and Cultural Change, 950–1350* (1993).
10. G. Phillips, *The Anglo-Scots Wars 1513–1550. A Military History* (Woodbridge, 1999), p. 41.

FIVE

European Overseas Expansion and the Military Revolution 1450–1815

Any consideration of the impact of the changes summarized as the Military Revolution in European expansion in the early-modern period requires an assessment both of the extent of this expansion and of the nature of the Military Revolution. This chapter considers alternative conceptualizations of European expansion in order to offer a critique of the dominant current notion of the global dimension of this Military Revolution. An approach that, instead, focuses on ideological factors is adopted. This approach, however, is interpreted flexibly. In considering and emphasizing the will to deploy resources and to conquer, it is necessary to relate attitudes to means and local opportunities, in order to avoid a misleading assumption that 'the West' could conquer whenever it was minded, and that it was all a matter of people making up their minds.

This chapter addresses the question of the nature and role of relative military capability in the long early-modern period, from the late fifteenth century to the close of the eighteenth. This question has frequently been discussed in terms of technology, with an emphasis on the role of firepower, although with stress also on clearly important developments in shipping and fortification. Thus, both European conquest and the Military Revolution have been presented as aspects of an early-modern gunpowder revolution, most profitably in the work of Geoffrey Parker. Far from concentrating simply on technology, Parker, like Michael Roberts, argued that weapons (technology) were important as they were marshalled by doctrine and the cultural habit of use (technique), and they drew connections between military technology and techniques and larger historical consequences.

Especially if the global geographical span of Parker, rather than the narrower European focus of Roberts, is adopted, then European overseas conquest can be seen as *the* Military Revolution. It is indeed possible to suggest that this process of conquest was revolutionary in cause, course and consequence, and was the aspect of military development that was most important on the global scale. Such an approach classically focuses on the arrival and establishment of Portuguese naval power in the Indian Ocean and the Spanish conquests of Aztec Mexico and Inca Peru, all of which were apparently able to meet these criteria. Their chronological near-congruence, in the period 1500–40, both dates this Military Revolution, or aspect of the military revolution, and ensures that arguments about the significance of any one of these events are supported by the presence of the others, the whole contributing to a sense of major change. This approach can be expanded to encompass changes in Christendom's apparent ability to fight the Turks, specifically the contrasts between the Christian failures of 1521–6 (the losses of Belgrade and Rhodes, and defeat at Mohacs) and the successes of 1565–71 (the relief of Malta, victory at Lepanto, and the length of time Szigetvár held out in 1566).[1]

The standard account of the Military Revolution, however, has been challenged by much of the research produced over the last decade.[2] There has been an emphasis on the variety of factors that contributed to European success, and the nature and extent of this success have also been qualified. The net effect is, first, to ensure that what occurred appears as more complex, and second, to throw doubt on established patterns of causality. This creates the problem of establishing a new synthesis, and of addressing the tension between complexity and the desire for a synthesis, or even simplification, both for its own sake and in order to provide a stage in military history.

Unless such a synthesis can be provided by military historians, there is a danger, first, that military history will excessively focus on detail, without the benefit of sufficient contextualization, and second, that dated and misleading syntheses will be offered. They will be provided either by non-military historians, who have a general tendency both to offer misleadingly structural accounts and to demilitarize military history, or by military scholars who fail to match new conceptualization to fresh empirical research.

The comparative dimension offers a fruitful approach towards a synthesis. It is useful to relate developments in one particular area to those in other regions, to consider land alongside naval warfare, and to provide a global context. All of these are also important tasks for those interested in warfare today.

Such an approach, however, faces serious problems. In particular, there is the danger that a comparative or contextual stance will lead to a hierarchy of military strength, capability and achievement based on the misleading idea that there was a commonality of challenge and an agreed basis that can be established and employed in the judgment of success. In addition, there is a risk that such an approach might lead to a teleology of military progress that is as misleading in its judgment of military success as it is crude in its understanding of governmental, social and cultural relationships. Instead, there is a need to understand that the plurality of military options and trajectories on the global scale reflected not some hierarchy of success in diffusion and acculturation, but instead, a complex process of adaptation to a multiplicity of circumstances.

Such a caveat may seem to undermine any process of contextualization, and the very complexity of the task, methodologically and in terms of material, both extent *and* lacunae, is forbidding. Nevertheless, such a process is still valuable. The practice of comparison is a reminder of military options that forecloses determinism. It adds point to understandings of contemporary debates. In the last half-millennium, the shrinking of the world (or looked at differently, the expansion of some of its regions) gave particular force to the issue of comparison, not least through the bringing together, in conflict, confrontation or co-operation, of hitherto separate military systems, and thus through a greater potential (and, in many cases, perceived need) for the diffusion of developments.

Military history is commonly teleological in character, and that helps to account for the popularity of the notion of the early-modern Military Revolution, but as indicated in the last chapter, this theory can be challenged. It is necessary, instead, to consider a later onset of 'modernity', and to look not only at the empirical case for the crucial changes as occurring later than the sixteenth century – whether, for example, with flintlock, bayonet and larger standing armies and navies in 1680–1730, or with the Revolutionary and Napoleonic period of 1792–1815, or with later, nineteenth-century, changes – but also at the wider intellectual, cultural, political and social contextualization of war and armed force.

Alongside contemporary suggestions of revolutionary change, it is worth noting that many of the accounts of conflict in the sixteenth, seventeenth and eighteenth centuries do not readily support the notion of a Military Revolution, understood in terms of a significant increase in European military effectiveness; although, of course, a characteristic of military commentators is that they complain about deficiencies, and of military equipment and forces that they are a wasting asset, especially in war. Major General John Richards recorded of the British siege of Valencia in the War of the Spanish Succession, in May 1705: 'our mortars could not be worst served . . . slow . . . they shot as ill as could be . . . the fuses were so ill made that . . . great many of them never burnt at all . . . length of time taking it has delayed us'. Moving on to Alburquerque, he was affected by the strength of the wall 'of a prodigious hard matter, armed with square towers after the ancient manner': an interesting comment on the continued employment of old methods.

Fifty years later, Brigadier General James, 2nd Lord Tyrawly, a veteran of the War of the Spanish Succession who was Britain's bumptuous envoy in Lisbon, reported on the deficiencies of the Portuguese and Spanish armies, then close to war, adding: 'I am confident ten thousand good dragoons drawn up on the frontiers between the two countries might take their choice which metropolis

[Lisbon or Madrid] they would march to, or perhaps a much smaller body.'³ This was the kind of fantasy that those confident of the effectiveness of their forces frequently resorted to (and still do). Ignorant of political context and consequences, such an approach was also frequently unfounded on a military basis, as Napoleon's forces were to discover in Spain in 1808. It was a state that was easy to invade, and a country that was impossible to conquer.

An alternative approach is to argue that many of the changes used to define the Military Revolution, or associated with it, such as larger armies, greater military expenditure, new tactics, and the *trace italienne*, all had medieval precedents.⁴ From this perspective, the major technological innovations in weaponry employed in the sixteenth century occurred earlier, in, or even prior to, the fifteenth century. There was only limited change in firearms technology during the sixteenth century, although an increase in numbers was important. The most sustained and imaginative criticism of the Parker thesis has indeed come from medievalists, although, in response, it can be suggested that their contribution really adds up to a prehistory of the Military Revolution. An analogy can be drawn with the argument that the Renaissance must be discussed with reference to the achievements of Carolingian (ninth-century), Ottonian (tenth-century) and/or twelfth-century Renaissances.

If long-term trends in the later Middle Ages are emphasized, it is unclear that subsequent European territorial and trans-oceanic expansion in any one period of, say, a half-century can either be linked to a unique stage in European military development, or profitably employed in order to focus on a specific period of European military activity. This is even more the case if the evolutionary theories of many recent historians of technology are employed as models,⁵ and there is an emphasis on users not designers, and on stability not breakthroughs; although Clifford Rogers has borrowed the notion of 'punctuated equilibrium', an approach that promises to combine both incremental and revolutionary change.⁶

It is important to clarify the linked questions of the putative Military Revolution and the military dimensions of European conquest. First, it is helpful to re-examine the components of the Military Revolution, in particular from the perspective of European overseas conquest. Here, one faces the problem of categorization, namely: which of the military changes in the period 1450–1660 are worthiest of note; whether, indeed, there was a staged or episodic early-modern Military Revolution, and if so, when it started and ended; and how far it is appropriate to think of a European Military Revolution that was different in type from military changes in other parts of the world. This is complicated by the need to consider whether to treat land and sea warfare separately or together.

As far as both land and sea are concerned, it is necessary to focus, not on initial 'discoveries', but on diffusion, the understanding and regularization of usage, and effective usage at that. From that perspective, it is pertinent to comment on the widespread dissemination of cannon and hand-held firearms in the sixteenth century. The cumulative firepower of European forces thus rose greatly, both on land and at sea, and both in Europe and overseas. These weapons could be transported and could be used in a variety of environments. A similar emphasis on sixteenth-century diffusion is appropriate for new-style fortifications, employing the *trace italienne*, which were essentially designed to thwart cannon. Whether these changes deserve the designation 'revolution', however, is unclear. For example, the effectiveness of new geometrical layouts in fortification is open to debate.

It is also unclear how best to evaluate the organizational changes of the period. It is uncertain whether terms such as 'bureaucratization' are appropriate, but the growing role of the European state gradually replaced the semi-independent military entrepreneurs of early days, especially from the mid-seventeenth century. Furthermore, Europeans were well advanced in the field of international finance, enabling states such as Spain in the sixteenth century, the United Provinces in the seventeenth and Britain in the eighteenth to finance their activities, in part, through a well-developed international credit network.

Aside from these questions, there is the issue of the effectiveness of the European military machine overseas, and the applicability there of the linked conceptualization of military change. It can be suggested that, even if there was a Military Revolution, it did not have a revolutionary effect overseas. It is possible to consider the Spanish *conquistadores* as unique,[7] and in addition, to regard the trajectory and causation of naval success, both in the Indian Ocean and elsewhere, as separate. Alongside these successes, there were numerous European failures, in the New World and the Indian Ocean, as well as in North, West and East Africa.[8]

It is important, moreover, to think of success and failure not only in European terms, but also in those of their opponents. The latter are frequently difficult to recover, but that does not make them less important. For the late nineteenth and twentieth centuries, it is possible to employ oral history, not least in order to probe, in their terms, what the opponents of Western powers thought they were doing. This approach is also relevant for conflict today.

Furthermore, in another limitation of European effectiveness, many opponents of European power were able to develop counter-measures, in weaponry, tactics and/or strategy. This was not a new process. The Arabs of the Middle East had succeeded in the twelfth and, more particularly, thirteenth centuries in adapting their tactics to counter the strengths of the Crusaders, who were, like their later European counterparts, also less numerous than their opponents, and heavily dependent on fortified positions.[9] It is interesting to contrast (Western) European expansion in the period 900–1250 with that in 1450–1815, and to consider why the latter was more successful, and why in some areas, such as North Africa in the sixteenth century, it was reversed. Parallels between the failure of the Crusading states and of Crusading attacks on North Africa, and the expulsion of the Portuguese and Spaniards from most of North Africa in the sixteenth century can be probed.

In the early-modern period, countermeasures by opponents of European forces included adopting similar weaponry and tactics, but also countering them, for example by fighting in more dispersed order and gaining cover, or by the use of such tactics as encirclement and feigned retreats. These tactics, however, should not be seen simply as countermeasures. Many had been employed for centuries. They became more appropriate in the face of improved firepower.

Neither the weaponry nor the organizational dimension of an early-modern European Military Revolution appeared as apparent a cause or aspect of relative capability in New England or Chile, West Africa or Mozambique, the Swahili coast of East Africa or the Persian Gulf, India or Sri Lanka, Java, Timor or the Amur Valley as might be implied by the literature on the revolution. It is all too easy to run together a few episodes and create a paradigm that can then be extrapolated – a problem that more generally affects the analysis of military history.

European failures qualify any account of success, but the fact that it was Europeans pressing against the indigenes of the far continents, and not vice versa, was important. In 1250, poor generalship at the battle of Mansourah cost the French opportunities of smashing Ayyubid power in Egypt; 270 years later, the then crucial overseas encounter for European power was taking place in Mexico. Outside pressure on Europe was only mounted by the Turks, a contiguous power, and by the naval forces of the North African Barbary states. The latter were largely an irritant, and a contrast between Barbary pressure on Europe and the European maritime impact in certain other parts of the world offers an instructive guide to European capability.

More generally, it is unclear whether an aggregate measure of achievement is more pertinent than a focus on individual failures or successes. Such a measure, whether or not used to support the notion of a Military Revolution, fails to address the multi-faceted nature of European expansion and military activity. It is dangerous to read from a particular success or failure in order to produce a general account of relative capability.

This is even more the case if the military activity of other expanding global powers outside Europe in this period are concerned. If European capability is discussed in terms of particular successes, and these successes then 'explained' accordingly, that cannot create a satisfactory explanatory model, because it fails to account for expansions such as that of the Ottomans, Safavids,

Mughals, Mongols (in the sixteenth century), Adal, Toungoo, and the Manchus.[10] For example, the great event in sixteenth-century India was the creation of the Mughal empire, not the expansion of Portuguese power,[11] which was very much limited to parts of the littoral, and was indeed challenged and, in part, reversed in the seventeenth century.[12]

In some cases, it is possible to accommodate successes by non-European powers, for example the successful invasion of the Ryuku islands by the Satsuma clan in 1609,[13] to the model of European expansion by focusing on similar features. Looked at differently, the 'Gunpowder Revolution' may have been important for the early Spanish conquests in the New World (which were achieved with *very* few Spaniards), but they were not in principle very different from other cases, both earlier and contemporaneous, of warriors conquering agricultural societies with much larger populations. Gunpowder weapons were employed in only some of these conquests. In short, gunpower was not a precondition for such conquests. It neither set the goal, nor was the agent that explains all.

It is unclear how far it is appropriate to distinguish between a European Military Revolution, in which new ways of war were invented (around the opportunity presented by firearms), and Ottoman, Mughal or other non-European use of firearms in accordance with William McNeill's model of gunpowder empires.[14] It can be argued that it was only in Europe that there was a sustained transformation of the culture of war. In contrast, in Ottoman Turkey and in Morocco, firearms were adopted as useful force-multipliers, as more or differently useful weapons, but the underlying culture of warfare remained the same, and this was even more the case in Persia, India and China. Such an approach is important if it is argued that the central issue of the Military Revolution is that of analysing and comparing European and non-European military institutions and cultures, rather than weapons. The effectiveness of any European transformation of the culture of war in the sixteenth century should not be exaggerated, however.

The gunpowder approach to early-modern military history is less commonly employed to focus on non-European powers, not least because, in their case, naval warfare cannot be readily brought in to support the analysis. The Europeans, in contrast, were more prominent at sea than on land. Seaborne 'plunder and trade' empires had existed before the Portuguese intrusion in the Indian Ocean, both there and in the Mediterranean. What was new in the Portuguese achievement was the vast distance from the base to the area of activity. Here guns, new sailing technology and ships built to fight with guns were all important, as guns did not require food and water on long expeditions and were not vulnerable to disease. Technological superiority facilitated, but did not determine, the consequences of both attack on and protection of limited objects: ships, forts and islands.[15]

Turning back to the non-European powers, the use of individual episodes of gunpowder success in order to construct a general theory of military effectiveness, capability and the rationale for change is as questionable outside Europe as it is in the case of the European states. In addition, in part, such assessments frequently rest on a failure to understand these episodes. Thus, for example, the oft-cited role of gunpowder firepower in the Ottoman victory over the Safavids at Chaldiran in 1514 is a matter of controversy. The Ottoman cannon, in part, were important because, once chained together, they formed a barrier to Safavid cavalry charges; in other words, they did not only offer firepower. More generally, it is unclear whether Ottoman firepower or numerical superiority was more significant at Chaldiran.[16] Similar points could be made about other battles and campaigns commonly discussed in terms of the successful use of gunpowder weaponry.

An important variable in any discussion that seeks to provide a hierarchy of weaponry effectiveness is provided by the quality of the firearms, which greatly affects comparability. Quality can be divided into, first, the refinement of the technology over time during the period 1350–1800 – in other words, the improvement of the basic models, and of the gunpowder – and second, the age of the firearms in the hands of the soldiers, and their maintenance. In the 1750s and 1760s, for example, many Indian mercenary troops came into the service of the British East India Company with their own weapons, which Company officers considered to be nearly worthless.

The evaluation of military success must not be dominated by Western criteria. Thus, rather than ignoring successful raiding by cavalry forces or treating it as a primitive form of conflict, such raiding can be seen as a corollary of European coastal positions, both serving as the basis for trade in a situation where trade and tribute were not polar opposites. The slave trades indicated this relationship. They served the requirements of more powerful societies, but were also dependent on a measure of local support. This was more apparent in the case of the Europeans in Africa[17] than of their Asian and Arab counterparts, but was also true of the latter.

Moving away from a concentration on gunpowder conflict does not have to extend to a denial of any role for such weaponry. In particular, it is possible to draw attention to two different, but probably related, developments: first, the increase in aggregate power flowing from the availability of such weaponry, and of troops that were trained to use it, and second, the training, systematization and industrial production that stemmed from the large-scale deployment of such weaponry on land and sea. The second development, in many respects, is the organizational definition of the Military Revolution. Such a definition is a cultural account that avoids some of the critique of technological determinism associated with the emphasis on weapons. Instead, there is a desire to understand the nature and context of weapons systems. As with the emphasis on weapons, however, it is unclear how far a process of optimization can be discerned, and also how it operated.

It can be argued that, from the sixteenth century, European forces acquired an edge in keeping cohesion and control in battle for longer than their adversaries, and that this permitted more sophisticated tactics in moving and withholding units on the battlefield and more effective fire. These techniques were developed in conflict between European forces. A British participant in James Wolfe's victory over the French outside Québec in 1759 recorded:

About 9 o'clock the French army had drawn up under the walls of the town, and advanced towards us briskly and in good order. We stood to receive them; they began their fire at a distance, we reserved ours, and as they came nearer fired on them by divisions, this did execution and seemed to check them a little, however they still advanced pretty quick, we increased our fire without altering our position, and, when they were within less than an hundred yards, gave them a full fire, fixed our bayonets, and under cover of the smoke the whole line charged.[18]

European forces tended to be able to advance, withdraw and retreat in a disciplined fashion. This ability can be linked to more general issues of administrative and political capability, can be related to the impact of a set of intellectual suppositions centred on Neostoicism,[19] and can be matched by a consideration of the deployment and employment of warships in battle.

This account has many strengths as a discussion of developments within Europe, but is less useful for any consideration of relative capability on the global scale, because it is not clear how to assess organizational effectiveness in a comparative context. In addition, there is a danger that any such assessment will rest on a primitivization of non-European traditions, which indeed ranged from Siberian tribes to Ottoman armies, and thus must have had very varied capabilities. Oriental forces, for example, appear as 'hordes', a Turkish word borrowed by Europeans better to understand the identity of a non-European warmaking culture, but one mistakenly employed to portray undifferentiated masses apparently unable to act in a planned fashion.

A revisionist approach should not be taken too far. It would be erroneous to suggest that the deep and real differences observable between European and Ottoman warmaking were simply a problem of European perception, an example of the 'Orientalism' later decried by Edward Said and others. To point out that Ottoman war culture never shrugged off some of the cultural values of the steppe, especially horsemanship and archery, is not to engage in a primitivization of that culture. Yet too little is known about Ottoman activities on the battlefield and on campaign to establish the case for any marked lack of organizational effectiveness on their part. It appears, at the very least, to be an exaggeration.[20]

115

Similarly, it is possible to offer a positive evaluation of the organization, conduct and effectiveness of African forces. In seventeenth- and eighteenth-century coastal West Africa, local forces were perfectly capable of organizing their armed forces in ways comparable to those of Europeans.[21] In Angola, in the sixteenth and seventeenth centuries, the Portuguese were successful only when supported by local troops.[22] Tactical and military-administrative developments unrelated to European models have been noted in other contexts, for example eighteenth-century Burma.[23]

Cultural and temperamental differences were important in both strategy and tactics, but there is no satisfactory methodology for the subject. One approach is Darwinian: warfare was such a competitive activity, sometimes a matter of life and death for states, dynasties and communities, that the military was arguably always driving to improve performance, even against tradition or fashion. The more continuous war is, the more powerful this factor is likely to be, especially with the dissemination of knowledge through the spread of military literature where methods and theories are tested against experience.

This approach can be employed to argue that Europe acquired a competitive advantage because of the frequency of war there in the period 1494–1815. Aside, however, from a lack of clarity as to whether the model is appropriate – namely, the question whether competitive 'evolutionary' pressures are indeed dominant in military culture and development, and if so, likely to lead to greater comparative capability – it is also unclear that Europe was more violent than, for example, India, Central Asia, West Africa, or China in the period 1600–1760. Similar socio-political pressures can also be seen in these areas, with the exception of China, where there was a measure of internal demilitarization. Rather than emphasizing bureaucratic purpose, the concept of 'social reproduction', an analysis suggesting that African states and warriors fought to sustain their position,[24] can also be applied to European states with the argument that absolutist societies were, by their nature, bellicose.[25] Furthermore, the notion that Western was different to non-Western warfare because the latter was less determined, uncompromising and

116

violent, and that its purpose was limited, even ritual,[26] is one that is based on a range of examples that may not be more generally applicable.[27]

The political corollary of the military primitivization of the 'non-West' is the argument about the frequent importance of disunity, factionalism and treachery among non-European peoples facing European conquest. This analysis does not relate only to the early-modern period. Instead, it can also be applied to earlier (and later) periods. For example, the Christian (re-)conquests of Sicily and Spain from the eleventh to the fifteenth centuries have been ascribed, at least in part, to Arab disunity, as has the longevity of the outnumbered Crusader states. This Arab disunity was also important in the subsequent Spanish and Portuguese advance into North Africa in the fifteenth and early sixteenth century, as indeed in fifteenth- and sixteenth-century operations against the Islamic khanates in what is now Russia and the Ukraine,[28] but, conversely, was not the case in conflict with the Ottoman Turks. In addition, such divisions were seen as important in subsequent Western expansion, for example the British conquest of India. The British military report on 'Arabistan' (South-West Persia) compiled by Air Headquarters, Iraq in 1924 claimed that 'owing to the absolute lack of cohesion amongst the tribes a general rising in the form of a single force operating under one control is not within the bounds of possibility'.[29]

In the early-modern period, disunity among non-European peoples has been seen as particularly important in India and North America, greatly helping European penetration, conquest and subsequent control, and also challenging any analysis in terms of 'the West versus the Rest'. However, the European colonial powers were themselves very divided, and also willing to arm and support the local opponents of their European rivals. They did so in both West Africa and North America in the seventeenth century, with the Portuguese and French abandoning initial refusals to sell firearms in response to the Dutch willingness to do so.[30] In 1741, Etienne de Silhouette, a French agent in Britain, reacted with alarm to the news that the British were arming Negroes in order to use them against

Spanish-ruled Cuba. He felt this might be very dangerous for all American Europeans, but argued that the British were too obsessed by their goals to consider the wider implications.[31] Munitions continued to be supplied regularly to non-European forces. In the late eighteenth century, Burma obtained cannon and muskets from British, French and Muslim suppliers. A century later, France and Russia provided Ethiopia with arms that it used successfully against Italy.

More generally, the standard 'cultural' approach to military difference suffers from the danger of hindsight, from reading back the successes of the West and the Westernization of military organization in parts of the world in the nineteenth century. Cultural issues are therefore not less important, nor the cultural approach less relevant, but the emphasis on organization and on other dimensions of military activity that can be considered in a cultural light is being asked to bear an excessive explanatory burden, and also cannot be employed with adequate precision.

It is unclear, for example, how far this analysis should be used even in the case of relations with the Turks, possibly the best-studied military 'interaction' with Europe, and one where a long time frame for inter-state conflict exists.[32] Alongside any emphasis on Turkish organizational obsolescence and its operational consequences, whether located in the wars with Austria of 1593–1606, 1663–4, 1683–99, or 1716–18, comes evidence of Turkish resilience and success, as in 1711–15 against Russia and Venice, and 1737–9 and 1788 against Austria, neither of which have been adequately studied.

Recent work emphasizes the continued strengths of the Turkish military machine into the seventeenth and even eighteenth centuries.[33] The Turks had not yet been written off. In 1715, Sir Robert Sutton, the British envoy in Constantinople, warned about the possible consequences of Turkish successes against the Venetian positions in Greece: 'If they are suffered to possess themselves of the Morea, now they are grown so powerful at sea and growing daily stronger, the Kingdoms of Naples and Sicily, as well as all Italy will lie greatly exposed to their insults.' He was wrong – their success had no such consequence – but his fears are worth noting. In 1732, the French envoy in Rome wrote to his counterpart in Vienna: 'La grande question est de savoir si les Turcs feront la guerre cette année au Chrestiens ou non.'[34] Eight years later, Villeneuve, the experienced French envoy in Constantinople, who had mediated the recent Treaty of Belgrade (1739) by which the Austrians returned Belgrade, Western Wallachia and northern Serbia to the victorious Turks, drew attention to Turkish superiority in the recent war, especially the size and logistical capability of their army.[35]

The Austrians were greatly helped in their conflicts and confrontations with other European powers in 1733–5, 1740–8, 1756–63 and much of the period 1792–1815 by Turkish passivity in Europe. Indeed, had the Turks attacked during these years, the situation might have been very dangerous for Austria. Austria itself was a conduit for a flow of military ideas that was both from East to West and from West to East. Poland occupied a similar position until the early eighteenth century.

It is possible that a similar re-evaluation emphasizing the strength and success of the opponents of European states may be pertinent for other military confrontations between European and non-European powers. In this context, it is probably mistaken to read between, and thus combine in the same analysis, wars with non-Europeans along Europe's land frontiers and those waged across the oceans. The latter, in particular, had an episodic character that reflected both the role of decisions whether or not to attempt landward expansion or control from coastal enclaves, and the dominance of essentially commercial roles. The central role of trade ensured that a set of values and relationships that did not focus on control or defence came into play.

This was far less the case with land frontiers. Issues of defence were often important in the latter case, certainly until the late seventeenth century, and they could lead to pressure for expansion. Thus, Tsar Ivan IV of Muscovy's conquest of the Islamic Khanate of Kazan in 1552 can be seen as a defensive response to the Khanate's alliance with the Crimean Khanate: 'Muscovy simply took over the frontier area to protect itself.'[36] On land frontiers, the Europeans were not generally able to offer the 'protection' that they could provide in coastal parts of South Asia.

Naval and, to a lesser extent, military skills were the major fields in which Europeans had an edge over South Asians up to the eighteenth century, and this made them useful allies. South and East Asia sold various high-quality manufactured goods that Europe could not produce. The trade was made up by European exports of (American) silver and 'protection'. However, the extent to which the Europeans had much to offer in land warfare prior to the development of sepoy forces in the mid-eighteenth century should not be exaggerated. In addition, there were important European setbacks in South Asia. For example, in the seventeenth century the Portuguese were driven from the Persian Gulf and Muscat, while the French failed in their attempt to intervene in Siam.[37]

Partial parallels between different relationships may still be drawn or considered. As far as both land and sea are concerned, it is striking how, after the initial period of European impact and expansion, which can be variously dated but may be centred on the period 1492–*c*.1560, there was a long stage during which, across much of the world, European expansion slowed or halted, or in some cases was reversed. This can be seen with the Portuguese in Angola, Mozambique and Morocco, particularly the last,[38] although expansion continued in Brazil. The Portuguese disaster at Azalquivir in Morocco in 1578 was not like the Italian disaster at Adua (Adowa) in 1898, a simple exception to the rule of European expansion. Furthermore, within forty years of 1898 the Italians had conquered first Libya and then Ethiopia. There was no comparable Portuguese expansion after Azalquivir. On Europe's land frontier, Ivan IV's conquests of Kazan and Astrakhan in the 1550s were not followed by any overrunning of the Ukraine or by war with the Crimean Tatars. The Crimea was not conquered until 1783. Russian expansion along the shores of the Caspian Sea and into the Caucasus was limited until the eighteenth century.

With both Spain and the Dutch, the slowing down came later than with the Portuguese. In the first case, there was a burst of continuing activity, especially in the 1560s and 1570s, for example in the

Philippines, and in the Mediterranean against the Turks, but thereafter this diminished. It might be attractive to suggest that this was due to a concentration of resources upon warfare within Christian Europe, an argument made for the second case by Fernand Braudel.[39] Yet such a concentration had also been true of the early decades of the century, when Spain had been heavily committed in the Italian Wars, but had still been able to project its power into the New World. Furthermore, the energy for expansion from the mid-sixteenth century, in large part, came from within the existing Spanish colonial possessions, Mexico leading to the Philippines,[40] and the latter to Taiwan, where the Spaniards established themselves in the 1620s. It is appropriate, instead, to draw attention both to a slackening in Spanish colonial expansion and to the strength of resistance, as in Chile, and to the north of Mexico, and on Mindanao.

In the case of the Dutch, later entrants on the colonial scene, expansion and attempted expansion in 1590–1670 was largely at the expense of other European powers, especially the Portuguese (although in Taiwan in 1642, the Spaniards),[41] and initial gains elsewhere, as on Java (Jayakerta/Batavia, 1619) and Cape Town (1652), did not lead to widespread expansion.[42] Again, this in part reflected a limited interest in territorialization. Dutch merchants did not seek to create Spanish-style *latifundia* (landed estates) across the world. The Dutch expedition sent in 1696–7 to explore the west coast of Australia reported that the 'Southland' offered little for the East India Company. The Dutch had been driven from Taiwan by Coxinga in 1662. They never re-established their position there.

Similar limits could be seen elsewhere, for example with the English on Sumatra and in West India, and all powers in West Africa. The French established bases on Madagascar in the 1660s and the 1740s, but did not conquer the island until 1894–5. In 1793, Captain John Hayes hoisted the British flag on the north-west coast of New Guinea, and, on behalf of George III, took possession of what he called 'New Albion', but the British Governor-General of India, Sir John Shore, and his council refused to support this private initiative, and in 1795 Fort Coronation and the colony were abandoned.[43]

This was a matter of political will as well as military capability. Both were important to conquest and resistance. They can be noted in the relatively small European forces sent to North America, Africa and South Asia.[44] The first centuries of the European presence in Asia were limited to seaborne activity, fortress-building and control of certain important islands; the Philippines, an extension of the Spanish empire in America, was, in part, a different case. There were no vast territories, huge agricultural populations or long land borders for the Europeans to defend in East or South Asia, although the situation in Siberia, much of which the Russians overran from the 1580s,[45] was different.

With superior technology, ships could defend themselves and help defend European island and littoral bases. The fortresses and their guns also represented a superior technology, but were seldom built in areas where they confronted major Asian rulers. When attacked, indeed, these fortresses were vulnerable. Parker devotes attention to Portuguese-held Malacca's success in fending off repeated attacks by the Sumatran Sultanate of Aceh, but, as he also notes, Muscat fell to the Omanis in 1650, Fort Zeelandia to Coxinga in 1662, and Mombasa to the Omanis in 1698. In 1686, when the Mughals vigorously pursued a dispute with the East India Company, the English evacuated their Bengal base of Hooghly and surrendered that of Bombay. They were only able to continue trading after apologizing and paying an indemnity.[46] Until the mid-eighteenth century, Europeans mainly defended themselves and their trade, and if they acted aggressively, did so, in part like nomadic raiders, against Asian mainland territories and Asian maritime trade which were taken under European protection.

Limited interest in conquest may, in part, be traced to the military factors discouraging expansion in the period *c.* 1560 to 1748.[47] Furthermore, it is necessary to consider the problems of conquest understood as the creation of new authority. This was less difficult where, as in Europe, there were existing political structures to take over or with which to reach an accommodation. It *could* be easier to force the more sophisticated, 'civilized' peoples into accepting new structures, just as it *could* be easier to defeat them.

In contrast, Amerindian resistance to the Europeans was at its most effective when it was 'primitive', especially if aided by difficult terrain or eco-systems, as in Amazonia and with the Araucanians of Chile.[48] Many such peoples practised dispersed warfare, which could frustrate more sophisticated, cohesive, concentrated European formations, as the Russians discovered against the Chukchi in north-eastern Siberia in the eighteenth century. Less urban market-based cultures could be harder to conquer. If peoples were nomadic, or with scattered or remote settlements, they presented the Europeans with fewer or poorer fixed assets to threaten. The British in India found this a fundamental problem in 'pacification' operations.

In such areas, there were also fewer political structures that could be acknowledged. As a consequence, a European presence could more readily lead to conquest, while, in the absence of recognition let alone mutuality, it was easier to justify the seizure and settlement of land.[49] This contrast could also be related to differences in the demographic relationship. In India it was easier, more profitable, and more necessary, to create a European presence short of conquest than was the case in North America, South Africa and Australia. The same contrast can be seen in the nineteenth century between British expansion in the more populous north island of New Zealand and in far less densely populated Australia.

There was a major difference between the establishment of a 'plunder and trade' presence, such as that of Portugal in the Indian Ocean, or even empire, and conquering and establishing authority over far-flung but lightly populated territories. In the first case, agricultural societies with many fixed assets were vulnerable to nomads and raiding warriors, such as Vikings, Mongols and Manchus. They could also be conquered by determined warrior groups, and such groups were able to offer protection against other possible assailants. The ability to protect created authority, and the Vikings, Mongols and Manchus understood this process. This mechanism also worked in the Spanish conquest of America. The Aztec and Inca empires rapidly fell, and the Spaniards took their role of rulers and protectors.

The conquest of extensive, lightly populated territories with nomads and 'primitive' people, such as North America, Australia and Kazakhstan, on the other hand, was a long-term project, largely achieved by emigration by a large number of Europeans. Across much of the world, this process was more pronounced in the nineteenth century than earlier, although it can be seen in operation in, for example, the Russian steppe lands in the seventeenth and eighteenth centuries. As already suggested, the establishment of an *intensive* structure of government and practice of power over more densely inhabited lands involved other problems. It was by its nature different both to the 'plunder and trade' empire and also to control over settlement colonies, and success proved far more elusive. *In part*, this helped to explain the obsolescence of empire in the twentieth century: imperial government increasingly entailed, as a goal, a degree of control that was incompatible with traditional practices of rulership.

In suggesting that a re-evaluation of the Military Revolution is necessary, it is appropriate not to go too far. A margin of European superiority in naval capability was maintained throughout the early-modern period, and this was also true of amphibious ability, especially the combination of ships, fortresses and garrisons. Non-Europeans were less successful in this sphere, although there were exceptions, such as the Turks in Rhodes in 1522 and Cyprus in 1570–1, and the Omani Arabs at Mombasa in 1698.[50] It is difficult to erode this European superiority by talk of other factors, such as the role of germs. Yet even in this sphere, it is possible to draw attention to the limited inshore effectiveness of European warships. Thus, in the seventeenth century, Western ship design was found to be deficient in operations against North African corsairs. This led the English to consider the use of galleys to defend their base at Tangier.[51] Furthermore, the unwillingness of the Chinese, Japanese and Koreans to maintain their powerful naval forces, which might have mounted a serious challenge to the Europeans, directs attention to the role of volitional factors, and thus to culture, identity, ideology, and politics. Choices were involved in investment in and sustaining navies: there was no determinism in the process.

Discussion of an early-modern European Military Revolution must note that, in contrast, the transformation in naval capability in the period 1820–1920 or 1855–1955 was greater than in any previous century; and this is true, more generally, of other aspects of military development. Advances in construction techniques and scientific developments had both been utilized earlier. The 1,095 foot-long ropery opened for the British navy at Portsmouth in 1776 may well have then been the longest building in the world. In 1788, Dr Charles Blagden visited the harbour works at Cherbourg, which he described as 'a new experiment in mechanics'. However, such experiments were more successful a century later, as, more generally, was the effectiveness of European government. Whereas the British naval bakery built in Portsea in 1724 could turn out 34 cwt of biscuits weekly, the new, larger factory opened in Gosport in 1828, with mass production equipment, was capable of producing 10,000 naval biscuits daily.[52]

From the seventeenth century, substantial navies were deployed by only a handful of non-European powers, principally in the East Indies, for example Aceh, the Mediterranean – the Barbary States of North Africa and the Ottoman empire – and in the western basin of the Indian Ocean – Persia, the Omani Arabs based on Muscat, and the Maratha Angria family on the Konkan coast of India. In the late seventeenth century, the Maratha fleet was considerable.[53] The ships of these non-European powers were longer in range than the war canoes of Madagascar, Oceania and West Africa, and more closely approximated to European warships, but they lacked the destructive power of the latter and the organizational sophistication of the European navies. The Barbary, Omani and Angria ships were essentially commerce raiders, rather than the more regimented and standardized fleets of the European navies, with their heavy, slow ships designed for battering power. The latter were also increasingly able to operate effectively and in sizeable numbers at a considerable distance from Europe. This owed much to the provision of local bases, such as Havana, and these were seen as important to European naval effectiveness. Thus, in 1788, the French Foreign Minister wanted to know if the British were constructing warships in Bombay.[54]

Furthermore, despite qualification of the concept of an early-modern Military Revolution, there were still important European land gains in the seventeenth and early eighteenth centuries. This was especially the case in Siberia, North America and Brazil. Such conquests were different to those made earlier at the expense of the Aztecs, Incas and, later, from the 1750s, at the expense of Indian powers, because the demographic balance in Siberia, North America and Brazil favoured the Europeans, especially at the point of contact, or was less unfavourable than it was before and after. That does not, however, minimize the importance of advances into areas such as Siberia, Canada, Louisiana and the thirteen colonies that were to become the United States.[55]

In addition, European projectors were eager to propose schemes for gains in many areas, reflecting a sense of confidence in the potential for Western expansion. Thus, in the early 1730s, a London merchant of Portuguese origin, John Da Costa, persuaded first a group of British merchants and nobles, next the Russian envoy, Prince Kantemir, and then an influential Russian minister, to support his scheme for a colony on a section of the Atlantic coast of South America allegedly unclaimed by any Atlantic European power. Da Costa proposed to gain the region with two warships and 500 troops.[56] British and Spanish diplomatic representations led to the end of the scheme, but in the same period the British established a colony in Georgia, and they were followed by Russian expansion across the Bering Sea into the Aleutian Islands and then Alaska. However, Portuguese projectors were less successful in 1635 and 1677 when they sought to advance up the Zambezi to gain the mines of Monomotapa.[57]

The period *c.* 1560 to 1748 can also be considered by contrasting it with the subsequent period, by assessing how far there was a 'tipping point' (critical moment of change) into the latter, and how far this was due to a case of more of the same, or how far to a new set of circumstances. In North America from 1748, there was no dramatic change, whether revolution or not, in terms of weaponry, but there was a new energy in European activity that essentially arose from competition between Britain and France. The same was also the case in India.[58] There was no particular increase in the quality gap, if any, in favour of European military methods, but more troops were deployed in both North America and India. Governor William Bull of South Carolina reported in 1761 that Anglo-American prisoners released by the Cherokees claimed that:

> their young men from their past observations express no very respectable opinion of our manner of fighting them, as, by our close order, we present a large object to their fire, and our platoons do little execution as the Indians are thinly scattered, and concealed behind bushes or trees; though they acknowledged our troops thereby show they are not afraid, and that our numbers would be formidable in open ground, where they will never give us an opportunity of engaging them.[59]

Anglo–French competition in the years immediately before the Seven Years' War (1756–63) and during the conflict itself were important for the projection of European power. Nevertheless, there was no common 'tipping point' in the mid-eighteenth century. In the case of the eastern Mediterranean, for example, although interest in the conquest of Egypt was expressed in 1739,[60] and then again from 1784, no attempt was mounted until 1798. The French were successful that year, their rifles 'like a boiling pot on a fierce fire',[61] but their conquest was brought to an end by British intervention. The British were far less successful when they landed in Egypt in 1807,[62] and Egypt was not conquered and held until the British invasion of 1882.

Further west, a Spanish attack on Algiers in 1775 was unsuccessful, and when, in 1786, Algiers signed a treaty promising to end piracy against Spanish shipping and to return Spanish captives, Spain had to pay 2.2 million pesos, including ½ million for redeeming the captives and 1 million as indemnification for the attack of 1775 and the bombardments of 1783–4.[63] In Burma and South-East Asia there was no major deployment of European force until the following century, in the case of Burma, the 1820s. French influence in Burma collapsed in 1757, when Pegu was occupied by Alaugpaya.

It is also worth noting that the difficulties that Europeans encountered, in the period of expansion from the late eighteenth century, in subduing and controlling far-flung territories involved ideological as well as operational issues. Aside from issues of conquest, there were grave problems in creating adequate political, organizational, commercial and financial rationales and patterns for control or influence.[64]

Interest in the conquest of far-flung positions and territories rose during the period 1792–1815.[65] The French Revolutionary and Napoleonic Wars led to another age of ambitious political 'projects', although the projectors were mostly officials of the imperial powers, particularly Britain. Although the conflict of that period involving European powers was concentrated in Europe, the French devoting a greater share of their military resources to conflict there than had been the case during the Seven Years' War, competition between the European powers again led to interest in the non-European world, encouraging, for example, Napoleon's invasion of Egypt in 1798, Russian expansion in the Caucasus, British and French diplomatic initiatives in Persia, and most significantly, British expansionism in India.[66] In 1812, Major General Charles Stevenson sent the Earl of Liverpool, Secretary for War and the Colonies, a memorandum urging the need to gain control of Timbuktu, a town only recently explored by Mungo Park:

Africa presents a new country and new channels for your industry and commerce, its soils favourable for your West India productions, it produces gums, drugs, cotton, indigo . . . gold . . . iron . . . this to England is infinitely of more consequence than the emancipation of South America . . . the teak wood so famous for ship building might be cultivated with success in some of its various soils . . . the possession of Thombuctoo [*sic*] would secure you the commerce of this quarter of the world and give you a strong check upon the Moorish powers of the Mediterranean by being able to intercept all their caravans and refusing them the commerce of the interior. It would likewise give you a complete knowledge of Africa to the borders of the Red Sea and to Ethiopia

. . . at the same time you could raise black armies for your East Indies and save your white troops for other service . . . [Napoleon knows] he can from the Niger pass his battalions in echellons to Cairo . . . not all the power of Great Britain can dislodge him if he first adopts the plan of establishing a chain of posts from the Upper Senegal to the Niger . . . I was not destined to conquer Africa, but bridle it, in order to have a check upon its Kings to protect British commerce as well as the African in its transit through the different kingdoms, by which means we should hold the country in check without the expense of defending it and by good management make the greater part of its sovereigns our friends, by supporting some, protecting others and augmenting their powers, and, as Allies, drawing from them whatever black battalions we may want.[67]

An emphasis on ideological issues that encouraged territorial expansion offers an approach to global military history that displaces attention from the technological interpretation, with its world conceived of as some type of isotropic surface. Politics was not simply a question of the mobilization of resources, but also one of the nature of a political society. This was important, both in terms of internal strength and of foreign policy. The ability of polities to incorporate insiders and outsiders varied (and varies), and was a crucial aspect of military strength. This was an aspect both of polities and of armies, the two not being separate in this context.

The Manchu state demonstrates the point. The successful integration of the Chinese and the Mongols in the Manchu state and army in the seventeenth century was crucial both to their conquest and to the subsequent success of the Manchu in conquering Tibet, Xianjiang and eastern Turkestan.[68] This paralleled the British conquest of much of India in 1757–1816, which was dependent on the integration of large numbers of Indians into the British military system.[69] Such a mixed military was fundamental to the imperial systems of the last millennium, especially before the dynamic nexus of the nineteenth century: namely, demographic expansion, nationalism, conscription and industrialization. The process of acquiring important political and military allies was most apparent with European imperial states operating across the oceans. It was also a factor, however, with powers operating against contiguous territories, such as the Russians in Kazan, Siberia, Central Asia and the Caucasus, as well as the Mughals and the Chinese. Local allies could offer necessary military skills. Thus, in the 1740s the British used Mohawks against pro-French Micmacs in Nova Scotia, while in 1742 Creeks helped the British block a Spanish advance on Savannah.

The ability to win and retain allies was not simply a matter of diplomatic success in specific conjunctures. It was also a product of a cultural flexibility that can be regarded as a crucial politico-military resource. The Manchu dynasty, for example, can in many respects be seen less as the government of China than as an imperial authority ruling China, Manchuria, Mongolia, Tibet, Korea and Xinjiang. The rulers consciously addressed themselves to different racial groups. Documents written in Chinese and Manchu, and originally thought to be simply translations, proved to say different things in the two languages. There was also an explicit use of Buddhism to control both the Tibetans and the Mongols. In Kashgar and Tibet, although not Xinjiang, the Chinese left government in the hands of the indigenous elite, and in Xinjiang, Qing (Manchu) authorities 'did not greatly interfere with local religion or customs'.[70]

Organizational capability and operational method were also important in the projection of power and the winning of distant allies. The creation of effective long-range logistical systems, such as those employed by the Turks to supply their forces, and by the Chinese in supporting their armies in Mongolia, Xianjiang and Tibet in operations from the 1680s to the 1750s, served to facilitate and sustain patterns of conquest, incorporation and alliance-creation that were very different to those of long-distance cavalry raiders. Indeed, it would be wrong to contrast the former with the European oceanic systems in order to suggest that the Europeans were necessarily more organized, even bureaucratic.

Effective logistics was not simply a matter of supplying forces. It was also important in winning the backing of local societies. Thus, the Turks rented, rather than seized, pack animals and wagons. Local support in frontier areas was also instrumental in the provision of accurate intelligence, which was very valuable in operations. Frontiers themselves were generally, although not always, porous zones, and an understanding of frontier societies and of concepts of the frontier throws much light both on expansionism and on warfare, on the goals as much as the methods of warfare, or, to reconceptualize the latter, the goals, and therefore the methods, of warfare.[71]

Characterizing the conquest of Xinjiang in the 1750s as a Chinese conquest of Turkic people is thus problematic, and Ross Hassig and John Thornton have made similar points about the Spanish conquest of the Aztecs[72] and Portuguese campaigns in Angola.[73] Much of the Chinese army was composed of Manchu and Mongol bannermen. The banner system enabled Mongols, Chinese and Manchus to work as part of a single military machine. As already suggested, it might be argued that the Manchu use of Chinese troops was much like the British use of *sepoys* in India.

Nevertheless, the degree of acculturation and assimilation of the Manchus into Chinese culture was greater, and therefore it is more appropriate to use the term 'China'. The contrast with the situation in India indicates the value of comparison. The strength of Manchu China owed a lot to the extent to which much of the territory that formed the initial Manchu homeland and the early acquisitions in eastern Mongolia had been the source of intractable problems for the previous Ming dynasty.

The overcoming of the frontier as a challenge to China was therefore as much a matter of political reconfiguration and reconceptualization as of specific military achievements.[74] There is, however, a danger that the notion of 'militarily advanced' and 'backward' currently defined in technological terms will be replaced by a similarly simplistic use of political concepts. Thus, there will be an overly simplistic emphasis on the ability of different state structures and political societies to create syncretic systems, maintain loyalty, mobilize military resources and sustain struggles, the whole constituting a form of political economy that has direct military consequences. Such an interpretation could be used to explain the success of expansionist states, for example Manchu China or Russia, at the expense of other political cultures, such as those, respectively, of Zungaria and of Kazan, Siberia, the Caucasus and Central Asia.

This approach is not without its problems, both empirical and methodological, as consideration both of the very varied political systems of the powers defeated by Revolutionary and Napoleonic France, and of their ability subsequently to triumph over France would suggest, but it does at least open the possibility of discussing the domestic background of war in a fashion that is at once systematic and theoretical. In particular, this approach focuses attention on the notion of political resources. However, it is also reliant on the creation of a deficient 'other', and this can lead to the notion that defeat in some way justifies its victim status. This is problematic from the moral perspective, and can also entail a failure to understand the nature of specific challenges. Powers, such as the Khanate of Kazan in 1552 or the Zungars in 1755–7 or the Kingdom of Poland in 1795, might be destroyed (by Russia, China, and the Russian-led partitioning powers of Russia, Prussia and Austria respectively), but that does not mean that they were bound to be so. The notion of political determinism in military history is as misleading as those of technological and economic determinism have been.

This can be seen if a counterfactual perspective is adopted. Counterfactualism can be employed by looking for alternative trajectories, such as demographic disaster due to disease for the European colonists in Latin America, East Asian naval development and expansion, or the defeat of the British in India by the local rulers, especially Mysore and the Marathas between 1780 and 1810. Counterfactualism can also be employed by applying a more complex critique to the cultural politics of conquest. As suggested in the previous chapter, the notion of an early-modern European world as a prelude to the modern Western world is an artificial and misleading construct that exaggerates aspects of modernity in Europe, gives them a false causal power, and underrates conservative social, cultural and intellectual patterns in the sixteenth and seventeenth centuries, presenting a false antithesis to the rest of the world.

Conservatism in the European world was indeed disrupted in the eighteenth century by the 'Enlightenment', but it was far from inevitable that the new ideas and governmental practices of the period would do more than shore up existing structures, as they did in Turkey. Indeed, the interactions of change, novelty, reform and revolution in the eighteenth century were far from fixed, and this remained the case until the utilitarian hegemony and self-consciously rationalizing drive of the mid-nineteenth century. An understanding of Europe as fluid in its development can be seen as crucial to any counterfactual discussion of her expansion. This approach is developed in Chapter Seven.

The greater pace of European activity in the second half of the eighteenth century had an organizational dimension, conflating political, cultural and military factors, in the shape of a far more substantial deployment of regular forces in North America and India, and also the development of *sepoy* forces trained to fight in the European manner in India. The latter was more important in the long term, because it led to a major and permanent increase in the armies available to the Europeans in South Asia, and reduced the extent of the need for multiple capability on the part of their home armed forces.

Furthermore, the degree of incorporation into Western models was an important organizational shift in South Asia. It is unclear whether it would necessarily have been less efficient to retain Indian priorities, especially in light cavalry. However, the British (and French) were largely dependent on allied forces for these, and this limited them politically, as allies had their own agenda. The role of such contingents has been minimized, because of the emphasis on *sepoys*, but that may have led to a misguided emphasis on 'Westernization' in organizational capability. The role of political choice indeed emerges in the failure of the British to create an Indian cavalry army. There were other factors, including the difficulty of obtaining horses on the east coast, the cost, approximately twice that of infantry man for man, and the intractability of Muslim horsemen, but an unwillingness to train Indians in better cavalry tendencies was also important, as was the preference of London for a defensive stance, rather than the aggressive possibilities offered by cavalry.[75]

As far as the impact of the *sepoys* was concerned, it may be that the demographic dimension was more significant than the organizational. *Sepoys* may be more remarkable, not in terms of an organizational capability gap with differently trained and equipped Indian forces, but as providers of the numbers without which the Europeans would have been overwhelmingly dependent on warships, artillery fortresses and allies, as they were in West Africa, Angola and Mozambique. This issue was more serious in South Asia than in North America, because of the nature of both the demographic balance with the native population and of the local environment. Disease was far more of a killer in India. As Colonel Gordon wrote in 1808, about the possibility of a campaign elsewhere in the Tropics: 'My fears on that subject are the climate, the climate, the climate!!'[76] Demographic factors also put far more pressure on European organizational capability. Thanks to their small numbers, the Europeans had to worry about how many men deserted or died of starvation. Reliability of payment and logistics were both therefore more important.

An absence of a military multiplier in a part of the world where the Europeans were under pressure could have serious consequences for their territorial impact. This was the case on the Swahili coast of East Africa, whence the Portuguese were driven by the Omani Arabs, and, for a long time, on that of West India. Such an emphasis can be countered by suggesting that the European powers in this period were not primarily concerned with territorial gain, and that it is therefore inappropriate as a measure of relative capability. This is also pertinent in the case of trans-oceanic conflict between European powers.

Differences in resource availability owed much to political and military choices in expenditure, allocation and use. The British were helped in their conquest of Canada in 1758–61 by the presence of more units than the French. This permitted recovery from defeat, and enabled the pursuit of more objectives. However, this imbalance in regular forces in North America was a product of choice. In 1755, the French had chosen to send a relatively small force to North America, whereas in 1756 they successfully invaded Minorca, in 1757 sent substantial forces into Germany, and in 1759 prepared to invade Britain.[77]

It is also necessary to employ this perspective of the impact of choice when considering relations, confrontations and conflicts between European and non-European powers, or indeed between the latter. Thus, for example, perceptions of Turkish military effectiveness in the late eighteenth century, not least *vis-à-vis* the Europeans, might have been affected had there been a major conflict with Persia that Turkey had won. That, however, might also have provoked a process of military adaptation in which the Turks focused on the Persian challenge,[78] rather than on that from European powers.

To the list of important major struggles that did not occur, for example that of China and Portugal in the sixteenth century, may be added others that could have arisen from much of the trans-oceanic European presence, especially in the Old World in the period 1650–1750. There were very few clashes with the Mughals, other Indian powers, or the rulers of South-East Asia (with the exception of Java). This situation did not completely cease in 1750, as can be seen by comparing the policies of British Governors-General of India. Under Sir John Shore, who was in India in 1793–8, there was an emphasis on trade, not territorial expansion. Shore used the army, under the Commander-in-Chief, Sir Alured Clarke, to depose and replace the ruler of Oudh, but he otherwise adopted a cautious and reactive approach, reasonably so, as the British position outside Bengal and the Carnatic was less strong than a concentration on these regions would suggest. The Maratha leaders, the Nizam of Hyderabad, and the Sikhs were able to develop their military power, and Shore did not take vigorous steps against Tipu Sultan of Mysore. The situation was to be very different under Shore's aggressive successor, Richard Wellesley, who launched major wars with Mysore in 1798 and the Marathas in 1803.

Such changes in policy were not restricted to European expansion. The death in 1723 of the Kangxi emperor was followed by a more limited Chinese policy towards both Tibet and Xianjiang under his less bellicose successor, the Yongzheng emperor (1723–36). The latter concentrated on domestic reforms, and was essentially a stabilizer.

When European overseas bases were not used to support major and sustained programmes of conquest, this can be explained functionally in terms of both major conflict and limited population growth within Europe. However, such functional and global explanations are of limited value. It is, instead, necessary to consider motivation and purpose. For example, issues of cost were important for the European presence in Africa and Asia, as was seen with debates inside the English and Dutch East India Companies on the question of the building and maintenance of fortresses during the later seventeenth and the early eighteenth centuries. We have been too focused on the quality (and quantity) of the means applied to major developments in European expansion, and military history, and not sufficiently focused on motives.

To close the chapter on such a note might appear unhelpful. There is no intention of offering a demilitarization of military history, or of ignoring the contingent, the conjunctural, and the operational dimension. Nevertheless, a cultural interpretation that focuses on reasons for conquest, or for the avoidance of aggressive warfare and conquest, rather than on technological or organizational enablers, directs attention to moods as well as moments. Such an interpretation also focuses on the methodological difficulties of devising a general theory of military capability and change in the early-modern world. An awareness of attitudes, diversity and difficulties is more appropriate than any simplistic and deterministic model that may be helpful to systems theorists of state-formation and the global economy, but inappropriate for scholars trying to understand the nature of military power.

NOTES

1. M. Roberts, *The Military Revolution, 1560–1660* (Belfast, 1956); G. Parker, *The Military Revolution: Military Innovation and the Rise of the West, 1500–1800* (2nd edn, Cambridge, 1996); T. Arnold, 'War in Sixteenth-century Europe: Revolution and Renaissance', in J.M. Black (ed.), *European Warfare 1450–1815* (1999), pp. 23–44.

2. J.M. Black, *War and the World: Military Power and the Fate of Continents 1450–2000* (New Haven, CT, 1998); W.R. Thompson, 'The Military Superiority Thesis and the Ascendancy of Western Eurasia in the World System', *Journal of World History*, 10 (1999), pp. 143–78.

3. Richards Diary, BL Stowe papers, vol. 467 fols 21–3; Tyrawly to Duke of Newcastle, 19 May 1735, PRO SP 89/38 f. 33.

4. A. Ayton and J.L. Price, 'The Military Revolution from a Medieval Perspective', in A. Ayton and J.L. Price (eds), *The Medieval Military Revolution: State, Society and Military Change in Early Modern Europe* (1998), p. 16. I have benefited from reading an unpublished paper by Kelly DeVries, 'Was there a Renaissance in Warfare? Humanism and Technological Determinism, 1300–1559'.

5. For an emphasis on an 'evolutionary' approach to the 'gunpowder revolution', see B.S. Hall, *Weapons and Warfare in Renaissance Europe* (Baltimore, MD, 1997).

6. C.J. Rogers, 'The Military Revolutions of the Hundred Years War', in C.J. Rogers (ed.), *The Military Revolution Debate: Readings on the Military Transformation of Early Modern Europe* (Boulder, CO, 1995), p. 77.

7. D.H. Peers (ed.), *Warfare and Empires: Contact and Conflict Between European and Non-European Military and Maritime Forces and Cultures* (Aldershot, 1997), p. xviii.

8. P. Powell, *Soldiers, Indians and Silver: The Northward Advance of New Spain, 1550–1600* (Berkeley, CA, 1952); R.C. Padden, 'Cultural Change and Military Resistance in Araucanian Chile, 1550–1730', *Southwestern Journal of Anthropology* (1957), pp. 103–21; M. Newitt, *Portuguese Settlement on the Zambezi* (1973); A. Hess, *The Forgotten Frontier: A History of the Sixteenth-century Ibero-African Frontier* (Chicago, IL, 1978); R. Law, '"Here is No Resisting the Country": The Realities of Power in Afro–European Relations of the West African "Slave Coast"', *Itinerario*, 18 (1994), pp. 51–2; J. Hemming, *Red Gold: The Conquest of the Brazilian Indians, 1500–1760* (2nd edn, 1995), pp. 72–3, 78–9, 90–6; R.G.S. Cooper, *Cross-Cultural Conflict Analysis: The 'Reality' of British Victory in the Second Anglo–Maratha War 1803–1805* (PhD., Cambridge, 1992).

9. C. Marshall, *Warfare in the Latin East, 1192–1291* (Cambridge, 1992).

10. J.M. Black (ed.), *War in the Early Modern World 1450–1815* (1999).

11. D.E. Streusand, *The Formation of the Mughal Empire* (Delhi, 1989).
12. E. Winius, 'Portugal's "Shadowy Empire in the Bay of Bengal"', *Camoes Center Quarterly*, 3, nos 1 and 2 (1991), pp. 40–1.
13. U. Suganuma, 'Sino–Liuqiu and Japanese–Liuqiu Relations in Early Modern Times', *Journal of Asian History*, 31 (1997), p. 53.
14. W.H. McNeill, *The Age of Gunpowder Empires 1450–1800* (Washington, DC, 1989).
15. C.M. Cipolla, *Guns and Sails and Empires: Technological Innovation and the Early Phases of European Expansion, 1400–1700* (1965); R.C. Smith, *Vanguard of Empire: Ships of Exploration in the Age of Columbus* (Oxford, 1993).
16. R. Savory, *Iran Under the Safavids* (Cambridge, 1980), pp. 41–4.
17. J. Thornton, *Warfare in Atlantic Africa 1450–1800* (1999).
18. Journal, possibly by Henry Fletcher, Providence, John Carter Brown Library, Codex Eng. 41.
19. P. Wilson, 'European Warfare 1450–1815', in J.M. Black (ed.), *War in the Early Modern World*, pp. 193–4.
20. C. Finkel, *The Administration of Warfare: Ottoman Campaigns in Hungary, 1593–1606* (Vienna, 1988); R. Murphey, *Ottoman Warfare 1500–1700* (1998), esp. pp. 85–129; G. Ágoston, 'Habsburgs and Ottomans: Defense, Military Change and Shifts in Power', *Turkish Studies Association Bulletin*, 22 (1998), pp. 126–41.
21. R.A. Kea, 'Firearms and Warfare on the Gold and Slave Coasts from the Sixteenth to the Nineteenth Centuries', *Journal of African History*, 12 (1971), pp. 185–213, and *Settlements, Trade, and Politics in the Seventeenth-century Gold Coast* (Baltimore, MD, 1982); R. Law, *The Oyo Empire c. 1600–c. 1836: A West African Imperialism in the Era of Atlantic Slave Trade* (Oxford, 1977), and 'Warfare on the West African Slave Coast, 1650–1850', in R.B. Ferguson and N.L. Whitehead (eds), *War in the Tribal Zone: Expanding States and Indigenous Warfare* (Santa Fe, New Mexico, 1992), pp. 103–26.
22. J. Thornton, 'The Art of War in Angola, 1575–1680', *Comparative Studies in Society and History*, 30 (1988), pp. 360–78.
23. V. Lieberman, 'Political Consolidation in Burma Under the early Konbaung Dynasty 1750–c. 1820', *Journal of Asian History*, 30 (1996), pp. 162, 168.
24. R.L. Roberts, 'Production and Reproduction of Warrior States: Segu Bambara and Segu Tokolor, c. 1712–1890', *International Journal of African Historical Studies*, 13 (1980), pp. 389–419, esp. 400–19.
25. J. Schümpeter, 'Zur Soziologie der Imperialismus', *Archiv für Sozialwissenschaft und Sozialpolitik*, 56 (1918–19); J.M. Black, *Why Wars Happen* (1998), pp. 32, 66–70, 98, 101–2.
26. McNeill, 'European Expansion, Power and Warfare since 1500', in J.A. de Moor and H.L. Wesseling (eds), *Imperialism and War: Essays on Colonial Wars*

in Asia and Africa (Leiden, 1989), p. 4; Parker, *Military Revolution*, pp. 118–19; J. Keegan, *History of Warfare* (1993), pp. 387–92; W.J. Eccles, *The Canadian Frontier 1534–1760* (New York, 1969), p. 6; P. Malone, *Indian and English Military Systems in New England in the Seventeenth Century* (Ann Arbor, MI, 1971), pp. 30–1; D.K. Richter, 'War and Culture: The Iroquois Experience', *William and Mary Quarterly*, 40 (1983), pp. 528–59; J. Forsyth, *A History of the Peoples of Siberia: Russia's North Asian Colony 1581–1990* (Cambridge, 1992), pp. 19, 51; A. Reid, *Europe and Southeast Asia: The Military Balance* (Townsville, Queensland, 1982), pp. 1, 5.

27. L.H. Keeley, *War Before Civilization: The Myth of the Peaceful Savage* (Oxford, 1966).

28. D. Ostrowski, *Muscovy and the Mongols: Cross-Cultural Influences on the Steppe Frontier, 1304–1589* (Cambridge, 1998).

29. PRO WO 33/1130, p. 102.

30. J.P. Puype, 'Dutch Firearms from Seventeenth Century Indian Sites', in J.P. Puype and M. van der Hoeven (eds), *The Arsenal of the World: The Dutch Arms Trade in the Seventeenth Century* (Amsterdam, 1996), pp. 52–61.

31. Silhouette to Amelot, French Foreign Minister, 7 September 1741, PRO 107/49. The letter was intercepted by the British.

32. G. David and P. Fodor, *Hungarian–Ottoman Military and Diplomatic Relations in the Age of Suleyman the Magnificent* (Budapest, 1994) is an important recent study. See also G. Ágoston, 'Gunpowder for the Sultan's Army: New Sources on the Supply of Gunpowder to the Ottoman Army in the Hungarian Campaigns of the Sixteenth and Seventeenth Centuries', *Turcica*, 25 (1993), pp. 75–96; and V. Aksan, *An Ottoman Statesman in War and Peace: Ahmed Resmi Efendi, 1700–1783* (Leiden, 1995).

33. R. Murphey, 'The Ottoman Resurgence in the Seventeenth-century Mediterranean: The Gamble and its Results', *Mediterranean Historical Review*, 8 (1993), pp. 198–200; G. Ágoston, 'Ottoman Artillery and European Military Technology in the Fifteenth to Seventeenth Centuries', *Acta Orientalia Academiae Scientiarum Hung*, 47 (1994), pp. 46–78.

34. Sutton to Sir Luke Schaub, 12 June 1715, New York Public Library, Hardwicke papers vol. 42; Polignac to Bussy, 16 February 1732, AE CP Autriche, supplement 11.

35. Villeneuve to Amelot, French Secretary of State for Foreign Affairs, 16 January 1740, Paris, Bibliothèque Nationale, Manuscrits Français 7191 fol. 5; Frederick William I of Prussia to Charles of Brunswick-Bevern, 24 August 1737, Wolfenbüttel, Staatsarchiv, 1 Alt 22 Nr. 609 fol. 164.

36. G. Ostrowski, *Muscovy and the Mongols*, pp. 187–8.

37. S. Subrahmanyam, *The Portuguese Empire in Asia 1500–1700* (Harlow, 1993), pp. 77–8; O. Prakash, *European Commercial Enterprise in Pre-Colonial India* (Cambridge, 1998), pp. 139–43. On Siam, see D. Van der Cruyse, *Louis XIV et*

le Siam (Paris, 1991); M. Jacq-Hergoualceh, 'La France et le Siam de 1680 à 1685: Histoire d'un échec', *Revue française d'histoire d'Outre-Mer* (1995), pp. 257–75.

38. W. Cook, *The Hundred Years War for Morocco: Gunpowder and the Military Revolution in the Early Modern Muslim World* (Boulder, CO, 1994).

39. F. Braudel, *The Mediterranean and the Mediterranean World in the Age of Philip II* (2nd edn, 1981).

40. J.L. Phelan, *The Hispanization of the Philippines: Spanish Aims and Filipino Responses, 1565–1700* (Madison, WI, 1967).

41. G.D. Winius, *The Fatal History of Portuguese Ceylon: Transition to Dutch Rule* (Cambridge, MA, 1971).

42. M. Ricklefs, 'Balance and Military Innovation in Seventeenth-century Java', *History Today*, 40 (1990), pp. 40–6.

43. On the weakness of the British in West Africa (only 75 effectives), Noailles, French envoy in London, to Vergennes, 27 February 1778, AE CP Ang. 528 f. 465; A. Griffin, 'London, Bengal, the China Trade and the Unfrequented Extremities of Asia: The East India Company's settlement in New Guinea, 1793–95', *British Library Journal*, 16 (1990), pp. 151–73.

44. The small numbers of troops available were also a factor elsewhere, for example in Central America, see W.S. Sorsby, 'The British Superintendency of the Mosquito Shore 1749–1787' (PhD, London, 1969), pp. 281, 285 re Spaniards in 1782.

45. G.A. Lantzeff and R.A. Pierce, *Eastward to Empire: Exploration and Conquest on the Russian Open Frontier, to 1750* (Montreal, 1973); T. Armstrong, *Yermak's Campaign in Siberia* (1975); Forsyth, *History of the Peoples of Siberia*, pp. 26–83.

46. B.P. Lenman, 'The East India Company and the Emperor Aurangazeb', *History Today*, 33 (1982), pp. 36–42. A lack of cannon dissuaded the Mosquito Indians from attacking a Spanish fort in 1762, see W.S. Sorsby, 'The British Superintendency of the Mosquito Shore', p. 126. However, for the limitations of fortification programmes, see pp. 132, 143, 145.

47. P.J. Marshall, 'Western Arms in Maritime Asia in the Early Phases of Expansion', *Modern Asian Studies*, 1 (1980), pp. 13–28. For the problem of insufficient settlers, in this case Spaniards in Texas, see R.S. Weddle, *The French Thorn: Rival Explorers in the Spanish Sea, 1682–1762* (College Station, TX, 1991), p. 302.

48. D. Sweet, 'Native Resistance in Eighteenth-century Amazonia: The "Abominable Muras" in War and Peace', *Radical History Review*, 50 (1970), pp. 467–81. For the difficulties that the European environment posed for invaders see, for example, D. Sinor, 'Horse and Pasture in Inner Asian History', *Oriens Extremus*, 19 (1972), pp. 181–2; R.P. Lindner, 'Nomadism, Horses and Huns', *Past and Present*, no. 92 (1981), pp. 14–15; J. Gommans, 'The Eurasian

Frontier After the First Millennium A.D.: Reflections Along the Fringe of Time and Space', *Medieval History Journal*, 1 (1998), p. 132.

49. W.E. Washburn, 'The Moral and Legal Justifications for Dispossessing the Indians', in J.M. Smith (ed.), *Seventeenth-Century America: Essays in Colonial History* (Chapel Hill, NC, 1959), pp. 24–32; A. Frost, 'New South Wales as *Terra Nullius*: The British Denial of Aboriginal Land Rights', *Historical Studies*, 19 (1981), pp. 513–23.

50. C.R. Boxer and C. de Azvedo, *Fort Jesus and the Portuguese in Mombasa 1593–1729* (1960).

51. A. Deshpande, 'Limitations of Military Technology: Naval Warfare on the West Coast, 1650–1800', *Economic and Political Weekly* (Bombay, 25 April 1992), pp. 902–3; D.F. Allen, 'Charles II, Louis XIV and the Order of Malta', *European History Quarterly*, 20 (1990), pp. 323–40.

52. Blagden to Lord Palmerston, 9 July 1788, New Haven, CT, Beinecke Library, Osborn Shelves C114; *Atlas of Portsmouth* (Portsmouth, 1975), section 3/6 (d). For the earlier deficiencies in attempts to improve techniques of fortification construction, see I. Coutenceau, 'Neuf-Brisach (1698–1705): la construction d'une place forte au début du XVIIIe siècle', *Revue Historique des Armées*, no. 171 (1988), p. 20.

53. M. Malgonkar, *Kanhoji Angrey, Maratha Admiral* (Bombay, 1959), p. 17.

54. Montmorin to Luzerne, envoy in London, 6 April 1788, AE CP Ang. 565 fol. 53.

55. See, for example, M. Giraud, *A History of French Louisiana* (5 vols, Eng. trans., Baton Rouge, LA, 1974–91)

56. L.A. Tambs, 'Anglo-Russian Enterprise Against Hispanic South America, 1732–1737', *Slavonic and East European Review*, 48 (1970), pp. 357–73.

57. R. Gray, 'Portuguese Musketeers on the Zambezi', *Journal of African History*, 12 (1972), pp. 531–3; M. Newitt, *Portuguese Settlement on the Zambezi* (1973), pp. 1–73.

58. The Indian National Trust for Art and Cultural Heritage, *Reminiscences: The French in India* (Delhi, 1997).

59. Bull to General Amherst, 15 April 1761, PRO CO.5/61 fol. 277, cited in S. Brumwell, *The British Soldier in the Americas, 1755–1763* (PhD, Leeds, 1998), p. 203.

60. F. Charles-Roux, *France, Egypte et Mer Rouge, de 1715 à 1798* (Cairo, 1951), pp. 1–2.

61. R.L. Tignor (ed.), *Napoleon in Egypt: Al Jabarti's Chronicle of the French Occupation, 1798* (Princeton, NJ, 1993), p. 37.

62. G. Douin and E.C. Fawtier-Jones, *L'Angleterre et L'Egypte: La Campagne de 1807* (Cairo, 1928); J. Dunn, 'All Raschid al-Kebir: Analysis of the British Defeats at Rosetta', paper read at the conference of the Consortium on Revolutionary Europe, Baton Rouge, LA, 1997.

63. J. Sabater Galindo, 'El Tratado de Paz Hispano–Argelino de 1786', *Cuadernos de Historia Moderna y Contemporánea*, 5 (1984), pp. 57–82.

64. For a recent example, see A. Webster, 'British Expansion in South-East Asia and the Role of Robert Farquhar, Lieutenant-Governor of Penang', *Journal of Imperial and Commonwealth History*, 23 (1995), pp. 1–25.

65. E. Ingram, *Commitment to Empire: Prophecies of the Great Game in Asia* (Oxford, 1981); S. Förster, *Die mächtigen Diener der East India Company: Ursachen und Hintergründe der Britischen Expansionspolitik in Südasien, 1793–1819* (Stuttgart, 1992).

66. D.M. Lang, *The Last Years of the Georgian Monarchy 1658–1832* (New York, 1957); M. Atkin, *Russia and Iran 1780–1828* (Minneapolis, MN, 1980); J.W. Strong, 'Russia's Plans for an Invasion of India in 1801', *Canadian Slavonic Papers*, 7 (1965), pp. 114–26.

67. Stevenson to Liverpool, 1 February 1812, Exeter, Devon County Record Office 152M/C1812/OF27.

68. F. Wakeman, *The Great Enterprise: The Manchu Reconstruction of Imperial Order in Seventeenth-century China* (2 vols, Berkeley, CA, 1985).

69. J.M. Black, *Britain as a Military Power 1688–1815* (1999), pp. 252–3.

70. J.A. Millward, *Beyond the Pass: Economy, Ethnicity and Empire in Qing Central Asia, 1759–1864* (Stanford, CA, 1998), p. 246.

71. P.C. Perdue, 'Military Mobilization in Seventeenth and Eighteenth-Century China, Russia and Mongolia', *Modern Asian Studies*, 30 (1996), pp. 757–93. The extensive literature on frontiers can be approached through D. Power and N. Standen (eds), *Frontiers in Question: Eurasian Borderlands, 700–1700* (1999).

72. R. Hassig, *Aztec Warfare, Mexico and the Spanish Conquest* (Harlow, 1994).

73. J. Thornton, 'The Art of War in Angola, 1575–1680', *Comparative Studies in Society and History*, 30 (1988), pp. 360–78.

74. T.J. Barfield, *The Perilous Frontier: Nomadic Empires and China 221 BC to AD 1757* (Oxford, 1989). For a less optimistic account, see L.A. Struve (ed.), *Voices from the Ming-Qing Cataclysm: China in Tigers' Jaws* (New Haven, CT, 1993).

75. G.J. Bryant, 'The Cavalry Problem in the Early British Indian Army, 1750–1785', *War in History*, 2 (1995), pp. 1–21, and 'Indigenous Mercenaries in the Service of European Imperialists: The Case of the Sepoys in the Early British Indian Army, 1750–1800', *ibid.*, 7 (2000), pp. 2–28; S. Alavi, *The Sepoys and the Company: Tradition and Transition in Northern India 1770–1830* (Delhi, 1995).

76. Gordon to General Craig, 7 May 1808, BL Add. 49512 fol. 17. He was referring to 'the Caraccas'.

77. A. Greer, *The People of New France* (Toronto, 1997), pp. 111–14.

78. For conflict earlier in the century, see R.W. Olson, *The Siege of Mosul and Ottoman–Persian Relations 1718–1743* (Bloomington, IN, 1975). For the different Safavid system, see R. Matthee, 'Unwalled Cities and Restless Nomads: Firearms and Artillery in Safavid Iran', *Pembroke Papers*, 4 (1996), pp. 389–416.

142

SIX

Politics, Resources and Conflict: Civil Wars, British and American 1639–1865

After earlier comments about the limitations of Eurocentricism, it might be considered surprising to devote a chapter to two civil wars that each occurred in Western powers. There are several reasons. First, they focus attention on the significance of civil wars and the use of force within states, themes outlined in the first three chapters, as well as permitting a comparison across time, and thus a probing of change.[1] Second, they develop the theme of counter-factualism within the West, a theme taken further in the next chapter. This underlines the need to assess the degree to which the identity of the West as a global force was neither fixed nor necessary. As a consequence, it is possible that the 'world question', that of the distribution and use of power within the world, might have been different: a different West might have been less determined or effective at extending its power. In short, political developments were crucial, and conflict within the leading Western powers was important to the relationship between the West and the 'rest'.

These points are important, although the issue of change could also be probed by considering non-Western examples, such as in China the revolts of Li Zicheng and Zhang Xianzhong in the1640s, and the Taiping revolution of 1851–64, and the, in part, associated Nian rebellion of 1852–68. Were there the basis for a systematic study, it would be worthwhile to compare these with the contemporaneous civil conflicts in Britain and America, but that is not possible. Hopefully, future researchers will take up the challenge. Instead, this chapter seeks to look at comparisons and contrasts between the British Civil Wars of 1639–52 and the American Civil War of 1861–5.

The mid-seventeenth century witnessed the greatest crisis in British society. Britons fought against and killed other Britons as never before. These conflicts are generally known as the English Civil Wars, but 'English' is wrong and misleading: the wars were about the control of the three kingdoms (Scotland, Ireland, England – the last including Wales), all of which were ruled by Charles I (1625–49).

Not only were there civil wars in England/Wales, Scotland and Ireland, but these were also inter-connected. In 1644, 25 per cent of serving troops were out of theatre – for example, there were 4,000 Irish troops in England and 6,000 in Scotland. And the wars were a major struggle. More than half the total number of battles ever fought on English soil involving more than 5,000 men were fought in 1642–51. Out of an English male population of about 1.5 million, over 80,000 died in combat and another 100,000 of other causes arising from the war, principally disease, a casualty rate far greater than that in any conflict Britain fought. This was also the case with the American Civil War. In no other war in American history was such a human and material effort made. The Confederate Army of Northern Virginia suffered a casualty rate of 20 per cent or more at each of the battles of Seven Days, Second Bull Run, Antietam and Chancellorsville, leading to 90,000–100,000 battle casualties in Lee's first year in command (1862–3). The Confederacy mobilized 80 per cent of its military-age whites, but by the spring of 1865, a quarter of the white Confederate manpool was dead and another quarter maimed.

Bitter civil conflict was not new in Britain, and more men may have fought in the Palm Sunday snow at Towton (1461) in the Wars of the Roses than in any of the battles of the Civil Wars, but the sustained level of hostilities in 1642–51, the Britain-wide scale of the conflict, and the vicious politicization of popular attitudes were unprecedented. The wars were vicious. After the Battle of Hopton Heath (19 March 1643), Sir John Gell, the Parliamentary Governor of Derby, paraded the naked corpse of the Royalist Earl of Northampton round the city. Few prisoners were taken by any side in Ireland. In 1649, when the Parliamentary New Model Army under Oliver Cromwell stormed Drogheda in Ireland, the garrison of about 2,500 was slaughtered, the few who received quarter being sent to work on the sugar plantations in Barbados. Largely as a result of subsequent famine, plague and emigration, the conquest of Ireland in 1649–52 led to the loss of about 40 per cent of the Irish population, and was followed by widespread expropriation of Catholic land.

The conflicts reflected Charles I's failure to run a multiple monarchy, indeed *any* monarchy, effectively. His inheritance from James I (and VI of Scotland) was a promising one, and, crucially, Britain was not involved in the Thirty Years' War (1618–48) on the Continent. Instead, it was well placed to compensate for its hitherto limited share in European overseas expansion. However, Charles lacked common sense, was devious and untrustworthy, and his belief in order and in the dignity of kingship led him to take a harsh attitude to disagreement. After encountering severe problems with Parliament over his financial expedients, Charles dispensed with it in 1629 and launched his Personal Rule. He was isolated from the wider political world, and his fiscal and ecclesiastical policies were unpopular.

Most people, however, were very opposed to rebellion, and it was a tribute to Charles' political incompetence that he transformed hostility into political disaster. The outbreak of civil war in England reflected a spiral of concern arising from Charles' mishandling of crises in Scotland (1637) and Ireland (1641). His commitment to religious change in Scotland and his aggressive treatment of Scottish views led to a hostile response. Instead of compromising with the resulting National Covenant (1638), Charles tried to suppress the Scots in the Bishops' Wars (1639–40). This was the start of the Civil Wars, and it was symptomatic of the whole period of conflict: Charles mishandled the situation and lost, religion played a major role in the war, and the conflict involved different parts of the British Isles, each of which were themselves divided.

To help deal with the crisis, Charles called Parliament in England. However, his period of Personal Rule had built up a series of grievances, and also much fear about his intentions. Previously loyal gentry turned against the King. In an atmosphere of mounting crisis, the need to raise an army to deal with a major Catholic rising in Ireland in 1641 polarized the situation. Who was to control this army? The role of this issue in crises helps explain why the control of the military was so important in many political systems. It was, indeed, the central issue in military history. Concerned about the situation, especially attacks on the ecclesiastical system, Charles resorted to violence, invading Parliament on 4 January 1642 in order to seize the 'Five Members', his leading opponents. They had already fled by river to the City of London, a centre of hostility to Charles. As both sides prepared for war, Charles left London in order to raise funds – a fatal step.

The resulting move towards war found the south and east of England, the navy, and many of the large towns backing Parliament, while Charles enjoyed greatest support in Wales, and north and west England. Although it is dangerous to adopt a crude socio-economic or geographical determinism, Parliamentary support was strongest in the most economically advanced regions, rather as that of the Union was to be in America in the 1860s.

Fighting in England started in Manchester in July 1642. There was a widespread desire to remain neutral, but, as later in America, in border states such as Missouri, the pressure of war wrecked such hopes, and the conflict spread. Both sides tried to raise the Trained Bands (militia), and initial moves by the opposing sides rapidly defined zones of control.

In the first major battle, Charles narrowly defeated the Earl of Essex, the uninspired general who commanded the main Parliamentary army, at Edgehill (23 October), but failed to follow up by driving decisively on London. He was checked at Turnham Green on 13 November. Charles failed to press home an advantage in what were difficult circumstances, and retreated to establish his headquarters at Oxford. His best chance of winning the war had passed. As the capital, London was an important source of legitimacy, as well as the major port and financial centre.

In 1643, the Royalists overran most of west England, crushing the Parliamentarians at Stratton and Roundway Down. Bristol, the major port on the west coast, fell to Royalist assault after a brief siege. However, the Royalist sieges of Gloucester and Hull were both unsuccessful, and the principal battle in the vital Thames Valley and surrounding area, the First Battle of Newbury (20 September), was inconclusive: the Royalist cavalry outfought their opponents, but their infantry were less successful. The Royalists had many successes in 1643, but did not challenge the Parliamentary heartland. The eleven mile-long defence system rapidly constructed for London – an earthen bank and ditch with a series of forts and batteries – was never tested in action, but was a testimony to the resources available for the Parliamentary cause.

As well as the major battles, there were also many small-scale actions that were important locally, for the war was both a national conflict and a series of local wars – a particular characteristic of civil wars, but one that is far less important in international struggles. At Winceby on 11 October 1643, for example, a minor cavalry engagement won by Cromwell, led to the Parliamentarians' capture of Lincoln, and was crucial to the course of the conflict in Lincolnshire. The war in south-west Wales swayed to and fro with only a limited relationship with the struggle elsewhere. More generally, local engagements reflected and affected the geography of the conflict, which itself led to the establishment of garrisons.

Communications were crucial to campaigning, logistics and the articulation of zones of dominance. Control over bridges, such as those at Upton-on-Severn and Pershore in the West Midlands and Tadcaster in Yorkshire, was very important in local campaigning. The bridge over the Avon at Stratford was broken by the Parliamentarians in December 1645, in order to cut communications between Oxford and the West Country. The county of Derbyshire was a linchpin of the Parliamentary cause in the Midlands, and therefore the goal of Royalist attacks from Yorkshire and the

Midlands. Control of Derbyshire, especially the crossings over the River Trent, would have linked these areas and helped the Royalists to apply greater pressure on the East Anglian-based Parliamentary Eastern Association. Royalist failure there lessened the co-ordination of their field armies, and led to longer lines of communication.

Thus, local struggles interacted with the national. The arrival of field armies could lead to the capture of positions that had conferred local control, as on 30 May 1645, when a Royalist force of over 10,000 stormed Leicester. Ably defended positions, however, could resist such pressure.

The relationship between local interests and national views is a particular problem for commanders in civil wars. Military units had a sense of locality, and many were reluctant to travel far. In 1645, the Northern Horse were able to return from Oxford to Yorkshire thanks to the threat of mutiny if they remained longer away from home and with the main Royalist field army. On the other hand, the Royalist Yorkshire Trained Bands moved into the Midlands in 1643, and the Earl of Derby had to send his best regiments from Lancashire to the main Royalist field army at Oxford that March.

Like international struggles, civil wars are greatly affected by the availability of foreign support. In January 1644, the Scots entered northern England on the side of Parliament. Both were united by opposition to Charles and suspicion of his religious leanings. Parliamentary successes in Yorkshire led the Marquess of Newcastle to pull back from Durham, and the Scots were able to press south and join the Parliamentarians in besieging York, the major city in northern England. The Royalist attempt to relieve it led, on nearby Marston Moor on 2 July 1644, to one of the two decisive battles of the war. The Parliamentarian/Scots army launched a surprise attack at about 7 p.m. The Royalist cavalry under Prince Rupert was successful on the allied right, but, on the allied left, Cromwell and Leslie drove their opponents' cavalry from the field. The infantry struggle in the centre ended when Cromwell's cavalry joined the assault on the Royalist infantry. Without hope of relief, York surrendered: the north of England had been lost for Charles I, as had any sense of Royalist invulnerability.

148

In the second decisive battle of the war – Naseby on 14 June 1645 – the newly organized Parliamentary New Model Army, under Sir Thomas Fairfax and Cromwell, defeated Charles, thanks in large part to the superior discipline of the Parliamentary cavalry. Prince Rupert swept the cavalry on the Parliamentary left from the field, but was unable to prevent his troops from dispersing to attack the Parliamentary baggage train. Cromwell, on the right, defeated the Royalist cavalry opposite and then turned on the veteran, but heavily outnumbered, Royalist infantry in the centre, which succumbed to an overwhelming attack. The leading Royalist field army had been destroyed. Thereafter, the Royalist situation was one of inexorable collapse. On 5 May 1646, Charles gave himself up to the Scots army in England.

The nationalization of the Parliamentary forces was important to the course of the conflict. The well-equipped New Model Army was a national army under the control of Parliament, not competing local interests, and able to operate in national strategic terms. It had a unified command under Fairfax, with Cromwell as commander of the cavalry. The army was more cohesive and better cared for than other forces, and its initial success was largely due to its being paid with remarkable regularity, at least for the first two years.

Parliamentary victory was due, in part, to the support of the wealthiest parts of England and Scotland, and to the folly of Charles, but chance also played a major role. The Royalist army in England was impressive, and in Scotland, Montrose was a good Royalist general. It took a long time for the Parliamentarians to create a winning team. The New Model's equipment and fighting style were essentially similar to those of their opponents; the major difference was that they were better disciplined and supported by a more effective infrastructure and supply system. Promotion was by merit, and Cromwell favoured officers and men imbued with equal religious fervour to his own. It became a force for political and religious radicalism, although the extent of this should not be exaggerated.[2]

The New Model Army reduced the significance of Scottish intervention, which had been very important in 1644. This intervention was one of the obvious contrasts between the British and American Civil Wars. It is, of course, difficult to determine the precise affects of Scottish intervention, but there is an obvious contrast with the American Civil War, which was unusual among major civil conflicts in that there was no external intervention. The contrast with the French Wars of Religion, the American War of Independence, the French civil conflict in the 1790s, and the Russian and Chinese Civil Wars is striking. Had, for example, Union victory or Confederate survival required British assistance, an assistance that would have been provided only on divisive terms, then the political consequences of the American Civil War would have been very different. In America, of course, there was no multiple statehood comparable to that in the British Isles in the 1640s. Indeed, that was a paradoxical benefit of the failure of the Americans to absorb Canada in the 1770s and 1810s.

Having won, the victors in the British Civil War fell out, with consequences for which there was no comparison in America in the 1860s. Parliament, the army leadership and the Scots clashed over Church government, negotiations with Charles, army pay arrears, and primacy. Agitators or delegates appointed by regiments pressed Parliament for arrears, and in August 1647 the army occupied London. The Scots were appalled by the rising influence of radical Protestantism in England, as opposed to Presbyterianism. In 1648, in the Second Civil War, the Scots invaded on behalf of Charles, who, in a major change of policy, had agreed to recognize Presbyterianism. In addition, there was a series of Royalist risings, particularly in Kent and South Wales. All were crushed, especially with Cromwell's victory over the overstretched and poorly co-ordinated invading Scots at Preston on 17 August. In contrast, when earlier allied with the Parliamentarians, the Scots had been able to campaign as far south as Hereford in 1645.

The army followed up by purging Parliament in order to stop it negotiating with the King (Pride's Purge, 6 December 1648), trying and executing Charles for treason against the people (30 January 1649), declaring a republic, and then conquering first Ireland and Scotland. Thanks, in part, to religious zeal, the army had become a radical force and had not been intimidated about confronting their anointed King. Parliament argued that Charles had given his word of honour not to fight again, and that he had broken it when he encouraged the Second Civil War. The army leaders were determined to punish Charles as a 'Man of Blood' who had killed the Lord's People. The execution made compromise with the Royalists highly unlikely, and entrenched the new ideological position of the new regime.

The Levellers, a radical group with much support in the army, pressed, in addition, for more extensive social and political changes, but their mutiny in the army in May 1649 was crushed by Cromwell. The Leveller rising was an early example of the potential radicalism of the military in civil war situations. Having invaded Ireland in 1649 and stormed Drogheda and Wexford, Cromwell attacked Scotland the following year, in response to the Scottish acceptance of Charles I's eldest son, Charles II. By the summer of 1652, all Scotland had fallen to the Parliamentarians.[3] It could no longer serve as an alternative model to English developments. The Scottish Parliament and Executive Council were abolished. The conquest of Scotland brought the period of conflict to an end, although Royalist conspiracies continued. Col John Saul led an unsuccessful rising in Norfolk in 1650, while John Penruddock seized Salisbury in March 1655, before retreating and being defeated at South Molton in Devon.

One of the major differences between the British and American Civil Wars was the political aftermath. In America there was a swifter return to civil peace, legitimacy, and unity. There was not only no powerful revanchist movement for Southern separatism, but also, in order to win, the Union had not embarked on a radicalizing process akin to that which eventually affected England in the Civil Wars, or France in 1793–4. The same was true of the American War of Independence, and is an important aspect of American exceptionalism. It is also a reminder of the very different consequences that can flow from civil war.

In Britain, in contrast, there was a protracted period of discontinuity and uncertainty. The regime of the Rump Parliament was bitterly divided, and in April 1653 Cromwell, now head of the army, closed the Rump. This made military rule blatantly clear. Royalist estates were sequestrated, and their owners permitted to have them back only if they paid a fine (compounded) and agreed to adhere to Presbyterianism. Royalist landowners also faced opposition from their tenants, including a refusal to pay rents and entry fines.

In 1655, authority in the localities was entrusted to Major-Generals, instructed to preserve security and create a godly and efficient kingdom – an unpopular step. The division of the country among the eleven Major-Generals created a new geography. It was a geography of military control that totally ignored long-established patterns based on aristocratic estates and clientages. The Civil Wars had revealed the military redundancy of the traditional centres, although they were of considerable local importance. Garrisoned stately homes, such as Burghley House, Compton Wynyates, Coughton Court, and the Marquess of Winchester's seat at Basing House, fell to siege or storming. The Major-Generals were instructed to keep control over the Justices of the Peace, the local gentry who had traditionally governed the localities, and to take charge of the militia. Compared to the former Lords Lieutenants, they lacked the local social weight to lend traditional strengths to their instructions.

An unpopular regime that was vulnerable to subversion and faced major financial problems, the Protectorate was nevertheless assertive abroad. Indeed, again in contrast with the American Civil War, the newly forged armed forces were employed abroad once they had no opponent at home. Benefiting from the opportunities created by conflict between France and Spain, war with Spain in 1655–9 led to the capture of Jamaica, but its expense caused a financial crisis. Thanks to larger armed forces, government was becoming more expensive and taxation more onerous. A major change in the nature of government was beginning.[4]

Cromwell died on 3 September 1658. His successor as Protector, his son Richard, unable to command authority and crucially lacking the support of the army, was deposed in 1659. The resulting political crisis, that saw the bitterly divided army opposed to the restored Rump Parliament which sought to end military rule, was only resolved with the Restoration. The army commander in Scotland, George Monck, marched south, occupied London, restored order, and was responsible for the election of a moderate Parliament that in 1660 recalled Charles II. Oliver Cromwell's corpse was exhumed and hanged at Tyburn.

The Restoration of the Stuarts did not, therefore, require another war, an outcome that would have been well nigh inconceivable over the previous decade. As a consequence, the discontinuity of the Republican episode is less obvious than would otherwise have been the case. Nevertheless, it was still apparent. Whereas in America rebellion was defeated in 1865, in Britain it led to a new political order. As such, it prefigured not the American Civil War, but what might be termed the 'Civil War within the Empire' or the 'First American Civil War', the struggle from 1775 to 1783 that led to American independence, and to a new state and society.

The unpopularity of the Interregnum regime encouraged postwar hostility to the notion of a standing army. This was important in affecting the nature of Britain as a military power. It would be misleading to attribute hostility simply to the Interregnum: the build-up of a standing army under the Catholic autocrat James II (1685–8) led to a similar reaction, while Britain's island identity encouraged navalism. Furthermore, it was still possible for British governments to field large forces on the Continent, as was to be shown in the War of the Spanish Succession (for Britain, 1702–13). Furthermore, weakness and failures in land operations on the Continent, for example in 1744–8, 1793–5 and 1799, helped to increase the perception of Britain as a naval power. They were important in the politics creating the strategic choice for what has been variously termed the Blue Water policy, the indirect approach, and the British way of war.[5] Yet it is also important to focus on political conjunctures, such as the Civil Wars and the Interregnum, in considering the development of attitudes that conditioned strategic cultures.

In both politics and conflict, it is appropriate to ask how far the civil wars in Britain and America can be seen as aspects of modernity. There is apparently a world of difference in technology and its consequences between the 'push of pike' of the infantry struggles in the British Civil Wars and the railways and telegraphs of the American Civil War. The railway made a major difference in strategy and logistics. It helped the North mobilize and direct its greatly superior demographic and economic resources, and played a major role in particular battles. Reinforcements arriving by train helped the Confederates win at First Manassas/Bull Run (1861). Rail junctions, such as Atlanta, Chattanooga, Corinth and Manassas, became strategically significant, and the object of operations. Man-made landscape features, such as embankments created for railways, played a part in battles. The North's dependence on railways led to the South raiding both them and the telegraph wires that were their counterparts.

On the battlefield, firepower, especially new, high-powered rifles such as the Springfield, proved more deadly than hitherto in North American conflict. The Model 1855, the standard infantry weapon in use in the American army in 1861, fired the Minié bullet and had a muzzle velocity of 950 feet per second. By 1863, much of the Union cavalry was equipped with breech-loading repeating carbines, which they used in clearing eastern Tennessee, and by 1865 some of their infantry had repeating rifles. The Gatling gun, an early hand-cranked, multi-barrelled machine-gun patented in 1862, was used in the war, following on from the use of the Colt machine-gun in the Mexican War of 1846–8. Even ordinary guns were far more effective than those available in the 1640s.

Due to defensive firepower, massed frontal attacks on prepared positions became more costly and unsuccessful, as the Union, in particular, discovered at Second Manassas (1862) and Fredericksburg (1862), and the Confederates at Corinth (1862), Stones River (1862–3), Gettysburg (1863) and Franklin (1864). The effectiveness of the defence was increased by the manoeuvrability of the well-armed riflemen. This permitted the ready presentation of new defensive fronts. Most of the casualties inflicted by rifle fire in the American Civil War resulted from long-range, accurately aimed defensive fire from behind entrenchments and log breastworks. Both

154

sides learned the necessity of throwing up entrenchments as a consequence of fighting each other to a costly draw at the battle of Antietam (1862). Bayonets and rifled muskets were increasingly supplemented by, or even downplayed in favour of, field fortifications and artillery – a sign of the future character of war between developed powers.

At sea, both sides deployed ironclads, as in the inconclusive duel between the *Monitor* and the *Merrimac* (renamed the *Virginia* by the Confederates) in Hampton Roads on 9 March 1862, the first clash between ironclads in history, and the manner in which both sides had ironclads when David Farragut's Union fleet successfully fought its way into Mobile Bay in August 1864. Also in 1864, the first effective attack by a submersible was mounted in Charleston Harbor, when the *Hunley* sank the Union screw sloop *Housatonic*, although she herself sank soon afterward, probably as a consequence of the force of the explosion. Later that year, the first successful torpedo boat attack occurred in Albemarle Sound, when, with a spar torpedo fitted to a launch, the Union sank the *Albemarle*, a Confederate armoured vessel similar to the *Merrimac*.

In contrast, a stress on technological capability and novelty is scarcely appropriate for the British Civil Wars or the American War of Independence. Neither struggle saw successful innovation (the first use of a submarine in 1776 was a failure), and neither was at the cutting edge of warfare in the period. Yet in the British Civil Wars, as in the American Civil War, there was a process of development under the pressure of conflict. The English forces in 1642 had no recent experience in conflict on land, and were less battleworthy than their Continental counterparts, who had been heavily involved in the Thirty Years' War, or indeed than the Scots, many of whom had participated as mercenaries in that war. Nevertheless, effective forces were created in the 1640s, not least the Royalist Western Army in 1643 and, especially, the New Model Army.

The same was true of America. The Mexican War (1846–8)[6] and policing operations against Native Americans were not an effective training ground for the Civil War. Many officers cut their teeth in the Mexican War, but it was different, both politically and militarily, to the Civil War. In 1861, the US army was only 15,000 strong and the navy had only about forty vessels in commission. Neither the Union nor the Confederacy were prepared for a major war in 1861: this was as true of the attitudes of their commanders as of the resources available. Yet, having been defeated that year at First Manassas/Bull Run, the Union reorganized its forces into the Army of the Potomac, developing a well-disciplined, well-equipped and large army. Similarly, the American (Union) navy turned from being a small force of deep-sea sailing ships to a far larger and more varied navy that was powered by steam and included many coastal gunboats able to mount a blockade and to support amphibious attacks. Ultimately, it became the second largest navy in the world.[7]

The Army of the Potomac showed, however, that organization and resources were of limited value without able leadership. George McClellan, who became General-in-Chief in November 1861, and who organized the new army, was indecisive, cautious and defence-minded, allowing the Confederates to gain the initiative in the eastern theatre for most of 1862–3, although his strategy was related to his belief in a limited war: 'It should not be at all a war upon population, but against armed forces and political organizations. Neither confiscation of property, political executions of persons, territorial organizations of state, or forcible abolition of slavery should be contemplated'.[8]

Ulysses Grant, in contrast, understood what it took to win a different war, a reminder of the need to match strategy to political purpose. He added a strategic purposefulness and impetus to Union military policy, and subordinated the individual battle to the repeated pressure of campaigning. The near-continuous nature of the conflict from his advance in May 1864, which led to the Battle of the Wilderness, combined with heavy casualties to give the war in the Virginia theatre an attritional character. This ground down the outnumbered South, which had a smaller army. Grant won the war. The South was outfought, defeated by military operations rather than by the weaknesses and contradictions of the Confederate war effort. Grant can be compared with Cromwell, not least because both subsequently gained power, although Grant (President, 1869–77) did so by constitutional means.

The importance of the Virginia theatre ensured that one apparent contrast between the British and American Civil Wars was less true in practice. The sphere of hostilities in Britain was far smaller than in America, but there was a degree of similarity. The two sets of capitals – London and Oxford, Washington and Richmond – were close, and much of the conflict in both wars took place within a hundred miles of one or other capital. In addition, there was campaigning in both wars that was more far-flung and fluid. Although entrenchments became more important in Virginia, further west and south the war was more mobile, as with W.T. Sherman's destructive 'March through Georgia' in November December 1864. In the case of the British Civil Wars, there was not the same intensity of conflict between Oxford and London, but much of the effort of the two sides did focus on attempting to put pressure on the capital of the other.

One major contrast relates to the British dimension. The wars in England, Ireland and Scotland were linked, but the politics of each was very different, and there were important contrasts in both military-political objectives and fighting techniques. There was no such comparison in the American Civil War. Instead, in so far as a comparison can be made, it should be with the War of Independence, for the war in Canada was very different to that in the Thirteen Colonies.

The process of teasing out similarities and contrasts is valuable, not least by throwing comparative light on particular facets of the wars. Yet there is also a wider question of development. The conflicts were more than two centuries apart, their outbreaks as far apart as the War of Independence is from us. They therefore offer an opportunity to probe the nature of military change. First, both conflicts remind us about the inherent symmetry of warfare within a military tradition. The two sides in both wars were more similar than either was, for example, to combatants in Asia in the period. It is tempting, and indeed important, to search for contrasts when focusing on individual conflicts, but it is the inherent similarities of organization, tactics and weaponry that are more apparent. They had been a feature of Western warfare since the spread of gunpowder weaponry in the early sixteenth century.

This similarity does not imply that there were not also important contrasts in leadership, resources and political-military objectives. Indeed, differences in resources and objectives could, and can, ensure that similar military systems operate in different ways, or that the fate of battle has very different meanings. The South was very much out-resourced, in manpower, economic strength and industrial production. The North had a 2.5 to 1 edge in manpower, and as many manufacturing establishments (110,000) as the South had manufacturing workers. In his Farewell Address to his soldiers in April 1865, the Confederate commander of the Army of Northern Virginia, Robert E. Lee, argued that they had been 'compelled to yield to overwhelming power': in short, beaten, not undermined by internal differences.

This disparity of resources was also apparent in the British Civil Wars and, as with them, invites the question of the inevitability of the eventual outcome. However, a move away from an assessment of military capability in terms of technology and resources and, instead, towards an enhanced awareness of political and military-cultural dimensions directs attention to developments that might have taken precedence over crude resource levels. It might have been impossible for the Parliamentarians and the Scots to agree terms in 1643. In this case, there would have been no Scottish invasion of northern England in 1644 and, arguably, no Royalist loss of Yorkshire later that year.

The role of contingency in the American Civil War is amply demonstrated with McClellan's advance on Richmond along the James River in 1862. Lee's success in blocking this cautious advance in the Seven Days battles (26 June–2 July), reversing the pattern of retreat set by his predecessor, Joseph Johnston, who was wounded on 31 May at the battle of Seven Pines,[9] arguably prevented a rapid close to the Civil War. A pattern of Southern defeat, that had also seen Nashville and New Orleans taken, most of Tennessee overrun and the *Virginia* scuttled, was broken. A rapid close to the conflict would have had political consequences, as it would have meant that the war would have ended before the emancipation of the slaves. McClellan might have won the next presidential election. Instead, his failure made it possible for the Republicans to attack him and for Lincoln to replace him, as part of the shift towards a more radical war with a different national (political) strategy in which civilian property was targeted. Although the North had a major lead in resources, the task of conquering the South, an area of ¾ million square miles, was formidable, and this posed questions for political and military will as well as strategy.[10]

The Seven Days began a series of Southern victories in the east that affected the military and political development of the war. Lee was a figure around whom the Confederates could rally. He understood that Confederate public opinion had a preference for taking the initiative, not responding to Northern moves, and that it sought offensive victories. The way he won was very important to public opinion,[11] – a point that has more general applicability than to the American Civil War.

Alternatively, war-weariness in the North in 1863–4 might have been stronger. This war-weariness could well have been encouraged had there been more Union setbacks. Already, in May 1863, Lee's victory at Chancellorsville (2–4 May) had helped the Peace Democrats. Northern war-weariness ensured that there was a need in the early summer of 1864 for an appearance of success. The heavy casualties in Grant's unsuccessful 'Overland' campaign and the initial failure to capture Atlanta hit civilian morale in July and August. In August, Lincoln feared that he would not be re-elected. However, a series of Northern successes in the Petersburg and the Atlanta campaigns and the Shenandoah Valley, including the occupation of Atlanta on 2 September, let Lincoln back into the White House.

To return to the question of military development. Aside from the issue of the continued importance of symmetrical warfare, it is apparent that in civil war the need to subjugate a hostile population remained important. Just as the Parliamentarians punished Royalists and Catholics with penal taxation, and the Patriot militia during the

War of Independence tyrannized the Loyalists, so the American Civil War witnessed an application of devastation as a means of conflict in a 'Western' context. Sherman considered he was fighting 'a hostile people', and set out to destroy the will of the civilian population, to 'make Georgia howl!' The result was $100 million worth of destruction. The Great Valley of Virginia was devastated in September 1864, South Carolina following in February 1865. The combination of an ability to crush organized resistance and to march unhindered across the Southern hinterland helped destroy civilian faith in the war.

Alongside continuities, it is also possible to observe contrasts. In the American Civil War, it was possible to offer a very different force–space ratio to that in the War of Independence (1775–83). Both sides in the 1860s could deploy large armies and take heavy casualties. More generally, demographic and economic expansion over the two hundred years from the mid-seventeenth century, particularly from the 1740s, transport improvements and different political attitudes ensured that there was not only a different resource base, but also a more optimistic attitude towards the procurement of fresh resources. The enhanced ability of the state to direct resources was joined to a view that it could and should.

The two combined in conscription. The Confederacy introduced conscription in 1862, the Union in 1863. Lee, a keen supporter of conscription, advocated the subordination of state rights to the Confederate cause, and believed, as a member of his staff testified, 'that since the whole duty of the nation would be war until independence should be secured, the whole nation should for the time be converted into an army, the producers to feed, and the soldiers to fight'.[12]

The South tried to control the economy and muster resources to a degree greater than that of the North. The utilization and organization of national resources, particularly those made available by new industrial processes, has been seen as an important aspect of military modernity;[13] although it has also been argued that the Civil War was more a continuation of the type of European conflict earlier in the century. This has been attributed to the difficulty of assimilating new technologies, and also the character of the generalship.[14] There was no general staff comparable to that being developed by the Prussians. The notion of war as a struggle between societies was not accepted by all. Conscription was unpopular with a large number of Northerners, while the Confederate forces were badly affected by desertion, although many deserters returned to the army, and it is important not to underrate the continued determination of the majority of the Southerners.[15]

The social practice of war also varied. Charles I essentially relied on traditional notions of honour, obligation and loyalty to raise troops. He epitomized the established social hierarchy, and his armies reflected this. Leadership for the Royalists was in large part a function of social position, although an increasing number of Royalist officers came from outside the social elite.[16] Aristocrats played a major role in the Parliamentary leadership in the early stages, but less so subsequently. The contrast between Prince Rupert and Oliver Cromwell, or between the Marquess of Newcastle and Sir Thomas Fairfax, was one of different attitudes towards responsibility, position, quality and merit. In this respect, the New Model prefigured the Continental Army of the American War of Independence, the Republican Army of the French Revolution, and the Red Army in the Russian Civil War. In each case, the radical force had to create and struggle towards a rationale for its actions that could survive early disappointments. In short, the army served as the expression of the political thrust of the revolution, as well as providing its force.

In the case of the American Civil War, the socio-political positioning was somewhat different. The Confederate army was the expression of Southern nationalism and, like the Continental Army, central to the proto-nation, but it was not a case of radicalism contained, as with the New Model, the French Revolutionary Armies and, albeit to a lesser extent, the Continental Army. Before it is argued that such forces were bound to prevail (and the Confederates conversely to fail), however, it is worth noting that each of the last could have lost.

The social and ideological politics of the New Model were important because they ensured that it was impossible to demobilize after the First Civil War. Here, the comparison is with the French Revolutionary Armies, not the Continental or Union forces. Indeed the postwar demobilizations of the latter two were crucial to American history (and arguably to the international history of their respective periods). This point returns us to the primacy of politics. Thanks to these demobilizations, there was a postwar in each case. Not only did the Americans not invade Canada as was feared, but the troops stationed in the south to support Reconstruction left in 1877 (and by then there were very few). Furthermore, northern politicians and public opinion came to support reconciliation. Lee received good obituaries in the northern press, and in 1898 the surviving Confederate generals were made US generals. Although there were four Black regiments in the post-Civil War army, there was no Black presence in the reconciliast view, much to the anger of Black veterans and leaders. Similarly, the Blacks who supported Britain in the Boer War (1899–1902) did not gain Boer farms and the vote, but instead suffered from the postwar search for reconciliation with the Afrikaners.

In the 1860s, American politics and society were not stamped with the mark of a permanent militarization, and, in the short term, there was a lack of promotion in the much reduced army, and an emphasis on peacetime professionalism.[17] In contrast, military rule was only ended in Britain by the implosion of the revolutionary regime in 1659–60, and in France by defeat and foreign invasion in 1814–15.

NOTES

1. For a comparison within the same period, see S. Forster and J. Nagler (eds), *On the Road to Total War: The American Civil War and the German Wars of Unification, 1861–1871* (Cambridge, 1996).
2. I. Gentles, *The New Model Army in England, Ireland and Scotland, 1645–1653* (Oxford, 1992).
3. For an effective recent survey, see M. Bennett, *The Civil Wars in Britain and Ireland, 1638–1651* (Oxford, 1997).
4. This theme is developed, specifically for the period 1639–74, by J.S. Wheeler, *The Making of a World Power: War and the Military Revolution in Seventeenth Century England* (Stroud, 1999).

5. P. Mackesy, 'Strategic Problems of the British War Effort', in H.T. Dickinson (ed.), *Britain and the French Revolution 1789–1815* (1989), p. 159.
6. On which, see in particular O.A. Singletary, *The Mexican War* (Chicago, IL, 1960), and K.J. Bauer, *The Mexican War, 1846–1848* (New York, 1974).
7. R.M. Browning, *From Cape Charles to Cape Fear: The North Atlantic Blockading Squadron During the Civil War* (Tuscaloosa, AL, 1993).
8. B.H. Reid, 'Rationality and Irrationality in Union Strategy, April 1861–March 1862', *War in History*, 1 (1994), p. 38.
9. S.W. Sears, *To the Gates of Richmond: The Peninsula Campaign* (New York, 1992).
10. P.A.C. Koistinen, *Beating Plowshares into Swords: The Political Economy of American Warfare, 1606–1865* (Lawrence, KS, 1996).
11. G.W. Gallagher, *The Confederate War* (Cambridge, MA, 1997), pp. 58–9, 115.
12. G.W. Gallagher, 'Another Look at the Generalship of R.E. Lee', in Gallagher (ed.), *Lee: The Soldier* (Lincoln, NB, 1996), 285.
13. E. Hagerman, *The American Civil War and the Origins of Modern Warfare: Ideas, Organization, and Field Command* (Bloomington, IN, 1988).
14. P. Griffith, *Battle Tactics of the Civil War* (New Haven, CT, 1989).
15. W. Blair, *Virginia's Private War: Feeding Body and Soul in the Confederacy, 1861–1865* (New York, 1998).
16. R. Hutton, *The Royalist War Effort* (2nd edn, 1999).
17. M.R. Grandstaff, 'Preserving the "Habits and Usages of War": William Tecumseh Sherman, Professional Reform, and the U.S. Army Officer Corps, 1865–1881, Revisited', *JMH*, 62 (1998), pp. 521–45.

SEVEN

Struggles in the Western 'Core': Counterfactual Scenarios 1688–1815

The military dimension of the 'Rise of the West' can be probed not only, as in Chapter Five, in terms of relations with the remainder of the world, but also, as in Chapter Six, by looking at the impact of rivalry and conflict within the West. This chapter focuses on the struggle between Britain and France in 1688–1815, although that is not the sole rivalry that is of importance, nor the sole struggle that could have resulted in a different character for the West. For example, the contingent events of 1776, 1914 and 1917 greatly altered the relationships between the USA, Germany, Russia and the rest of the West, and changed the political culture of the Western world. More generally, it is possible to stress a number of different courses, all of which would have altered the character of the West, and definitely, probably or possibly its impact in the wider world. Yet as it was the struggle or rivalry between Britain and France that was most important to the course of Western overseas expansion in the period 1680–1900, the 'tipping period' in relations between the West and Asia, Australasia, Africa and, arguably, North America, it is appropriate to focus on this struggle.

In doing so, it is possible to link five counterfactuals. Would a different course and result to that struggle have affected the relationship between the 'West and the Rest', both then and thereafter? Did Britain have to win? Could Britain have developed differently in this period? Could the Jacobites have won (one of the more obvious instances of the last, although not the only one: radicals, for example, could have been more successful in the 1790s and 1800s)? Could the British have suppressed the American Revolution?

These questions are very different in character and scale, and yet are all linked, which itself poses problems in terms of actual and implied relationship and causality, as well as of prioritization. As effects ramify through the system, so positive and negative 'feedback loops' greatly complicate causal and counterfactual inference in history.[1]

The explicit awareness of the role of the scholar as interpreter is an important aspect of counterfactual studies. This approach draws on the task of the historian in organizing material and asking questions, but takes it a stage further. As such, it is appropriate to add the authorial voice, rather like Henry Fielding in his novel *Tom Jones* (1749). By so doing, the role of the scholar as organizer of questions is made explicit and the counterfactualism can be introduced and discussed in a helpful manner. In addition, this process underlines a postmodernist dimension to counterfactualism, not that that is the sole philosophy/methodology that is at issue. However, a stress on postmodernism helps counter the charge that those interested in counterfactualism are primarily reactionaries reluctant to accept the process or results of change. This is particularly important in the military sphere, not least because counterfactual discussion is frequently associated with such a regret, for example over Southern defeat in the American Civil War.

How and why did the West triumph? Many answers have been given. In the nineteenth century, they focused on issues of cultural and religious superiority. Technological advances were seen as indicators of cultural norms – a concept revisited in a different form in the late twentieth century. For several decades, accounts of the rise of the West have primarily focused on technological explanations. At one level, this approach has concentrated on particular devices and their development. This has led to important work on the so-called Gunpowder Revolution, and comparable studies of aspects of maritime technology. More generally, technology has been seen as relating not only to devices, but also to their use as part of a culture of organization in which analysis, planning and informed prediction came to play a major role. In short, technology was the opposite of fatalism, it was an earned providence that was then used to construct rationality and a scientific world view in which dominance by the West was both natural and attainable. This view, furthermore, survived to affect subsequent analysis of the very process.

Yet the West was not only more varied than this notion might suggest. There was also greater complexity in what might be termed the cultural politics of conquest or, more bluntly, the 'why' question; and this question can be related back to the issue of the inherent variety of the West: in short, its problematic character. The culture of Europe was important to its global role. European dominance did not have to take the forms it did. In many respects, a distinctive form(s) did not arise until quite late.

It is possible to disaggregate Europe, and compare and contrast the Atlantic West and the Continental powers. Continental expansion by Austria, and especially Russia, was not so different to expansion by non-European empires, particularly once the initial 'mercantile' (Cossack) phase of the conquest of Siberia ended. Thus, a model of Euro-Asian landward conquest can be provided, and it can be suggested that it was not similar to maritime expansion.

Aside from disaggregating the West, this distinction provides another sphere for counterfactual analysis. It is possible that different political developments within Europe would have affected the balance between maritime and Continental expansion. These might have included the suppression of the Dutch Revolt by Philip II of Spain in the 1570s or 1580s, as well as the later struggle between Britain and France. It is also necessary to consider whether domestic developments within individual European states would have made a major difference. These might have included different political trajectories within France, such as the triumph of the Huguenots in the French Wars of Religion in 1560–1600, or in Britain, for example Stuart victory in the Civil Wars of the 1640s, or in Germany the earlier adoption of primogeniture, which would have

166

avoided the patchwork of tiny states, or more specifically, a success by the Habsburg Holy Roman Emperor Charles V in sustaining his victorious position after his victory over his North German Protestant opponents at the battle of Mühlberg in 1547. Mentioning these both highlights the number and range of indeterminacies and underlines the role of authorial choice. A focus on the Anglo–French struggle and on related conflicts within the British world reflects a belief that the period was of particular importance, and that options regarding European expansion had not been foreclosed by the earlier rise, first, of the Iberian powers, and, later in the seventeenth century, of the Dutch.

The threat posed by Jacobitism (the cause of the exiled Stuarts) to the British political and constitutional settlement of 1688–9 directs attention to the role of conflict in moulding the development of Britain, and thus of British relations with other Western powers, specifically France, and therefore, of the character of the West and its relationship with the 'rest'. The Jacobite challenge also raises important issues about the nature and evaluation of military progress, specifically 'modernization'. It is worth probing in some detail because it provides a case study of wider applicability that can be thus treated.

The Jacobite threat was most acute in 1745. That year, the Jacobite rising in Scotland under Charles Edward Stuart (Bonnie Prince Charlie) quickly succeeded, with a total victory at Prestonpans on 21 September. His opponents then assembled an army, under the elderly Field Marshal George Wade, at Newcastle. Wade was familiar with northern Britain, but his combat experience was on the Continent, and he proved too slow-moving to cope with Charles Edward. The latter avoided Wade by invading England via Carlisle, which fell after a short siege (10–15 November). The defences were not impressive, but the defending force was anyway insufficient in number and lacked civilian support. Wade's slow attempt to march via Hexham to Carlisle's relief was hampered by the winter weather, leading to a 'non-battle', a basic building block in counterfactual studies. Instead, the Jacobites advanced unopposed through Penrith, Lancaster, Preston and Manchester, *en route* towards London. Their advance west of the Pennines followed, at least initially, the course of the invading Scots in the Second Civil War in 1648 and that of the Jacobites in 1715, but, in contrast to both the former occasions, the advancing force outmarched its opponents. The ability of armies to respond differently to similar strategic parameters, and the role of leadership in their doing so, were amply demonstrated in these three advances.

Aside from that under Wade, another regular army was assembled to confront the Jacobites in the West Midlands. Commanded by George II's younger son, William, Duke of Cumberland, this force was out-manoeuvred, as a result of being misled by deliberately circulated reports that the Jacobites intended to advance on North Wales. Cumberland thus failed to stop a Jacobite advance on Derby, which was entered, without opposition, on 4 December. At that point, the Jacobites held the strategic initiative and were also in a central position, while their opponent's forces were divided. Wade's army was still in Yorkshire, and Cumberland's was exhausted by its marches. The government was assembling a new army on Finchley Common to protect London, but it was a relatively small force. Unlike the Scots, the English population had had no access to arms, and even the Whig militia, which did obtain arms in 1745, would not fight, perhaps less because they would not uphold the Hanoverian succession than because they did not regard themselves as soldiers.

Nevertheless, the Highland chiefs were disappointed by the lack of the support promised them by Charles Edward: both assistance from English Jacobites and the absence of a French landing in southern England. After bitter debates, the chiefs forced Edward to turn back, and he began his retreat on 6 December. The outnumbered and outgunned Jacobites were to be crushed by Cumberland at Culloden on 16 April 1746. Was this defeat inevitable?[2]

Culloden is seen as a defeat for an army emphasizing speed, mobility and primal shock power by a force reliant on the concentrated firepower of disciplined infantry and their supporting artillery. Visual force is lent by the contrast between rows of British troops with their standard weaponry, and the more fluid, but also archaic, individualism of their opponents, generally, but misleadingly, presented as without guns.

The mental picture created is somewhat similar to the modern film *Zulu* (1964), an account of the British victory at Rourke's Drift in 1879. It is made clear that the two sides are starkly different, and that one represents order and other characteristics commonly associated with military activity and progress. Indeed, although the Zulu charge was at times effective, it was only so at the cost of heavy casualties, and British firepower generally defeated it.[3] Thus, the Jacobites can be seen as representatives of an earlier form of warfare that was bound to fail, because it was redundant, for tactical and technological reasons. The Jacobites can be compared to other eighteenth-century forces using non-Western tactics, for example the Cossacks and the Tatars. They can be seen as examples of 'barbarian' warfare, rather as Edward Gibbon presented the Tatars,[4] or, at the very least, as dated by the March of Modernity, as an Ossianic fragment of times past, primitive virtue, and prowess that might be impressive on an individual basis, but was unsuccessful as an organized (i.e., social) means of waging war.

Another dimension can be added by noting the success of the British state in dealing with its military opponents in the British Isles from the 1640s. The New Model Army was successful under Cromwell against both Irish and Scots, so that, from the early 1650s, for the first time, all of the British Isles was under effective control from London. The failure of this to occur earlier indicates the questionable nature of what might be termed geopolitical determinism, not least because the weaponry employed in the 1650s was not dramatically different to that used with far less success by English forces in the mid-sixteenth century. In 1685, James II defeated the Duke of Monmouth's rising, and in 1688 he would have destroyed any domestic risings had it not been for the invasion of a regular army under William of Orange. William's forces suppressed opposition in both Scotland and Ireland in 1689–92, Jacobite risings in 1715–16, 1719 and 1745–6 were defeated, and in 1798 the Irish rising was crushed. Indeed, Scotland and Ireland both became major military resources for the British state.[5] Within the British world, only the American colonists were successful in rebelling, but the American War of Revolution (1775–83) is commonly seen as different, and treated, in both its military and political aspects, as a harbinger of a new world, an aspect of the Modernity of the period.[6] The apparent differences with the Jacobite risings thus help to underline the necessary failure of the latter.

Such an approach, however, can be challenged by suggesting both that Jacobite warfare was not anachronistic, and that the Jacobites came near to success in 1745. The entire set of assumptions underlying the supposed context and chronology of military modernity can be queried.[7] Detailed consideration and contextualization of the warfare of the period provides an empirical underpinning for the counterfactual argument. Although irregular warfare was atypical in Europe in the middle of the eighteenth century, refusing to follow suit in the rigorous orders that were imposed on warfare under the auspices of linear tactics was not specific to the Jacobites, but a general aspect of the 'little war' at the time. Furthermore, later in the century, there was to be a major breach in the rules pertaining to linear tactics with the formation of columns. Thus, the Jacobites can be seen as anticipating the kinds of warfare that were to become more prominent towards the end of the eighteenth century.

The Jacobites were helped by the unfortified nature of most of the British Isles. The major fortified British positions were naval dockyards or overseas bases, such as Gibraltar and Fort St Philip on Minorca. There was no system of citadels protecting major domestic centres of government, especially in England. Not only did this ensure that the Jacobites did not have to fight their way through a series of positions, losing time and manpower as they did so, but it also meant that the British army lacked a network of bases that could provide shelter and replenish supplies. After Charles Edward captured poorly fortified Carlisle, he faced no fortified positions on his chosen route to London. Thus, the apparent determinism offered by stress on fortifications as a sign of military development in the early-modern period, and its ability to reflect a hierarchy of capability based on resources and applied science, can again be subverted by an empirical understanding that opens the way for the counterfactual argument.

The Jacobites were, of course, different as a military force to the British state in several major respects. They lacked a navy, and were therefore unable to challenge British maritime strength, a strength that was crucial in the '45, especially because it covered the movement of British troops back from the Low Countries in 1745 and the supply of advancing British forces in eastern Scotland in 1746. The Jacobite army was also a newly created volunteer and non-regular force, with non-bureaucratic supply and recruitment systems, and this necessarily affected its *modus operandi*, not least in matters of control and command, and logistics. The extent of new voluntary recruitment was important, not least as an indication of political support.

Government action against Catholics and suspected Jacobites in England limited the popular support the rising might otherwise have received in England. There was no popular rising, but then there had been none that was decisive when William III invaded in 1688. A measure of English support was shown by lack of resistance, especially at Carlisle, where the aldermen presented the keys to the city to Charles Edward on bended knees. The English Jacobite leaders were waiting for the French to land, and had not agreed to rise until then.

In his *Decline and Fall of the Roman Empire* (1776–88), Edward Gibbon addressed the question whether 'barbarian' invasions of Europe could recur, deciding that advances in military technology, specifically the development of artillery and fortifications, made this highly unlikely. Thus progress, in the form of science, apparently allowed advanced and settled societies to employ a military technology that multiplied the impact of their soldiery. This he saw as crucial in enabling such societies to restrict military service to specialized forces, and thus free the rest of society for productive activity and civilized pursuits.[8] Progress therefore secured and benefited from military modernity – an analysis that could apparently be applied in Europe and then on the global scale to limit any possible alternatives.

The Jacobite forces, indeed Celtic warfare in general, were recognizably different in type to contemporary and subsequent concepts of military modernity, but the '45 proved Gibbon wrong. The more 'advanced' society in conventional terms was nearly overthrown, and its eventual victory was far from inevitable. The Jacobites won two of the three battles they fought in 1745–6, and the projected night attack on Cumberland's forces at Culloden might also have been successful.

Even without French intervention, the British were badly handicapped in 1745 and 1775–83 by other factors that do not relate to the issue of whether the Jacobites had a regressive and the Americans a progressive military system. The Jacobites were more of a threat because they could readily threaten the centres of British power, but equally, relative propinquity ensured that they could easily be attacked. This interacted with the issue of time, one of the central factors in counterfactual arguments. The British failure to crush the American rebellion in 1775 gave the Americans time to organize themselves politically and militarily, to extend the rebellion greatly, and to weaken the Loyalists. The Jacobites lacked this margin, just as they lacked the wealth of the American economy; both were arguably more important than the location of each movement in terms of military modernization.

Had Charles Edward marched south on 6 December 1745 . . . is a question that can be matched by a number of hypotheticals for the War of Independence. These include: what if the British fleet had moved into the East River to block Washington's retreat from Long Island in August 1776, Clinton had advanced on Philadelphia in December 1776, as he suggested, Howe had moved north from New York in 1777, France had not entered the war, or there had been more British ships off the Chesapeake in 1781? None of these were implausible.

The War of Independence was a series of highly contingent events, and this must undermine notions of inevitable American victory. Even the American folklore of the conflict, with its stress on the heroic character of the resistance at Bunker Hill in 1775, still more the crossing of the Delaware in December 1776, and the privations and training at Valley Forge in the winter of 1777–8, affirms contingency. This is appropriate. A close reading of American military correspondence, for example the papers of George Washington or Nathanael Greene, scarcely suggests a confident sense of victory. Instead, there is reiterated stress on acute deficiencies in resources, particularly manpower, money and munitions, and Congress found it difficult to create effective national institutions. There was also a need to respond to British advances that could not readily be pre-empted, particularly because of the strength of the British navy until 1778 (and indeed, also after French entry), and the extent to which, from 1776, New York and Canada were safe bases from which the British could mount offensives.

The political dimension is also important. Because America became independent, it is too readily assumed that Loyalism could only be a failed option for Americans. The political counterfactuals of the struggle are not probed, not least the possibility that it might have led to a different political solution, one in which negotiations produced a conclusion short of full independence.[9]

Furthermore, the issue of timing does not attract sufficient attention. In the Seven Years'/French and Indian War, it was crucial to British success that the war did not end in late 1757 or early 1758, or even in October 1759, when Pitt told the Prussian envoy that he was ready for a peace that did not involve retaining Louisbourg, or Québec if captured.[10] In the American War, it is not clear whether a longer conflict might not have led to a different result, especially if France had made peace after its naval defeat at the Saintes (Saints) in 1782. In the French Revolutionary and Napoleonic Wars, Britain was to make a greater and more sustained effort.

French-supported Stuart success in 1745–6 might have made the struggle for American independence much more different. Thus, the British 1745 counterfactual leads to an American counterfactual, one that challenges the benign exceptionalism that reigns in American public culture. Rather than seeing an independent America as necessarily a society with a federal state, weak government, a balanced political system and a relatively tolerant public culture, it is possible to envisage a process of radicalism and authoritarianism under the pressure of struggle and war akin to that of France in the 1790s. This would have affected the character of what might still have been the greatest twentieth-century power. It might have ensured that, as with France, and later Russia, more effort would have been devoted to exporting the revolutionary struggle. In particular, support from France and her ally Spain for Jacobite rule of America could have led an independent America to sponsor revolution in the West Indies and Latin America. This might have been a co-operative process, but as with France in the 1790s and 1800s, the result could as easily have been conquest and expropriation. Thus, a more authoritarian America might have become an imperial power in its own hemisphere, *possibly* further accentuating its domestic authoritarianism. This would have led to a stronger military, rather than the small army and poorly trained and equipped militia that was generally so unsuccessful in the war of 1812–15 with Britain, and the navy that lacked significant powerful long-range capability.[11]

The dynamics of the struggle might also have led to a very different America geographically;[12] just as postwar American dynamism helped extend the new state to the Gulf of Mexico.[13] Newfoundland might have joined a revolution against Stuart rule in 1746, Nova Scotia, but not Québec, such a revolution had it been held after 1763. There might also have been more energy devoted to extending the revolution to the British colonies in the West Indies, and even to Ireland.

Aside from the consequences for America, such a struggle, a bitter civil war *within* the British empire and thus the West, might well have left fewer resources and less drive for the projection of British power elsewhere, especially in India and to Australia. The British might have failed to subjugate the Marathas, Mysore and later the Sikhs with long-term consequences for the history of South Asia. The notion that there was a 'turn to the east' in British imperialism as a consequence of the loss of America is controversial, but not without value. It depended, however, in part, on the availability of resources.

Yet a Stuart Britain may not have been a French client state or pro-French for long. In 1678, Charles II had turned against Louis XIV. James II was neither the ally nor the client of Louis in 1688. While in exile at St.-Germain, he became so, but had he been restored, he would not necessarily have remained compliant. Charles Edward kept up an independent attitude to the French which worried them, and he would have been no tool of France. After all, the American colonists owed their independence to the help of France, but did not become clients of France – far from it. In contrast, the Hanoverian kings, George I and II, were allied to France in 1716–31, and accused by critics of being overly pro-French.

Having addressed the role of contingency as opposed to structure, and of force as opposed to other political considerations, in the case of Britain, it is sensible to turn to its principal opponent, France. The question why France failed is one that takes note of the sense of contingency that was, and is, so important in, and to, the discussion of international relations and warfare. In a period of war, international relations and warfare were merged, especially by the Revolutionaries and Napoleon, and it is important not to separate them. Any analysis of conflict that places an emphasis on the importance of the reasons why wars broke out and were sustained, and on the role of alliances, necessarily directs attention to the political dimension.

In 1750, France had been a major power in Europe, and one of the two leading European states in North America and India. By 1762 this was no more. The surrender to the British of Montreal in 1760 and of Pondicherry in 1761 brought the end of New France (Canada) and of French India. Defeat at Rossbach in 1757 and at Minden in 1759, at the hands of Prussia and an Anglo-German army respectively, shattered French military prestige in Europe and made it clear that during the Seven Years' War, France would be unable to settle the affairs of Western Europe, and thus also unable to play a major role in affecting developments further east.

At least in 1759–63 it was not metropolitan France itself that was occupied or that suffered territorial losses. The situation was very different in 1815. The Second Treaty of Paris of 20 November 1815, the diplomatic liquidation of Napoleonic France, stipulated an occupation of north-eastern France for five years, a large indemnity of 700 million francs, and the cession of important border positions. By the Quadruple Alliance of the same date, the four great powers – Austria, Britain, Prussia and Russia – renewed their anti-French alliance for twenty years, a step designed to limit the chances of France disrupting the alliance. Napoleon had failed, totally. Louis XIV had not had to accept terms anywhere near as harsh at Utrecht in 1713. Nor had France in 1788-90, weak as her international position then was during the early stages of the Revolution, been obliged to accept humiliations and setbacks of this order.

These were only the most conspicuous French failures. There had been many others. During the last quarter-century of the *ancien régime* the most apparent were the repeated failures to protect European allies, especially Poland and Turkey, and also, in 1787, the Dutch Patriots. Poland and Turkey lost heavily in 1772 (First Partition of Poland) and 1774 (end of Russo–Turkish war) respectively, and thereafter their situation continued to deteriorate, much to the concern of Vergennes, Foreign Minister in 1774–87.

The fall of the Dutch Patriots was an aspect of the failure to take full advantage of the War of American Independence (1778–83 for France). This conflict had created an impressive but unstable alliance, and the French were unable to sustain it into the postwar world – a crucial aspect in judging the success of any war. Aside from the loss of Dutch support, there was also a failure to retain that of the Americans, while relations with another wartime ally, Spain, cooled. In addition, the war itself had gone less well than had been hoped. The British had not been driven from India, Gibraltar or Jamaica, and in 1782 they had won the battle of the Saintes (Saints), the last major naval engagement of the war in the western hemisphere, and the only naval battle in the conflict that could be described as decisive.

Similarly, there were serious disappointments in the years of Revolutionary enthusiasm. These were both diplomatic and military. In the first case, the inability to win the support, or at least prevent the enmity, of Prussia, Britain and the Dutch in 1792–3, as had been hoped, was most significant. It ensured that the war would not be defined by the Revolutionaries as the short and victorious conflict with Austria they sought. Had it been so limited, the French might well have been successful earlier, with many consequences in terms of the domestic stabilization of the Revolution. There was no inherent reason why war with Austria should have lasted longer than the time it took for the Russians to subjugate Poland (1793–5), especially as Poland provided alternative opportunities for the Austrians to make gains. Danton offered terms to Prussia, Sardinia, Switzerland and Tuscany in an unsuccessful attempt to weaken the relative position of Austria and Britain. He sought a negotiated peace with Britain in late 1793, but French objectives were scarcely those that would satisfy her or Austria.

The international situation was transformed by the victories of the Revolutionary armies in Belgium in 1794, especially at Fleurus, but that prevented a serious reconsideration of French foreign policy that might have brought an end to the war. A more restricted conflict would have ensured less support for domestic opponents of the Revolution, and this might have discouraged opposition by Royalists, Federalists and others. The 1790s, in short, are the decade of contingency *par excellence*. There might have been no Terror, no creation of the Sister Republics, no government by the Directory, no Napoleon.

The Revolutionary French won some crucial and well-known victories, but also faced a series of defeats, such as Neerwinden in 1793. In combination, these ensured that the war lasted far longer than had been envisaged, or than France could afford without serious damage to its economy and to its relations with occupied areas. Individual failures left Britain dominant at sea, in Ireland, and in Egypt. More generally, the Revolutionaries encountered the classic problem of ideologies – the unwillingness of the world to conform to their ideas.

Having seized power in 1799, Napoleon had considerable success, both military and political. His rule was accepted in much of Europe, and many states were obliged or willing to co-operate. This had direct military value, not least in the provision of substantial allied forces for Napoleon's campaigns.[14] Yet this made the failure all the more apparent. In the funeral pyre of 1812–14, alliances were destroyed and Napoleonic rulers and officials driven away. The crippling failure that British naval might had already visited on the overseas French empire (first in the 1790s, and then again in the 1800s) was, thanks to the victorious co-operation of Russia, Prussia and Austria, repeated in Europe.

So France failed. It was not alone in failure, as any consideration of Spain in 1807–24, and to a lesser extent Britain in 1775–83 and Austria in 1787–1809, would reveal. Yet, there is a French failure that requires discussion. It throws light on the nature of French achievements, on the character of the French state, and on the warfare and international system of the period.[15] There is also a possible instructive parallel with modern consideration of an RMA and its consequences, because the methods (and objectives) of war, and military organization of the French Revolutionaries[16] can be seen as just such an RMA. New concepts were implemented, new capabilities exploited, and a national army created. This leads to the question of why this RMA did not lead to a long-term increase in French strength.

The first problem is one of classification. How is the failure we are discussing to be atomized? Are we to see, within French history, separate trajectories with different explanations, in particular a maritime struggle with Britain and a Continental struggle, or are these to be combined, and if so, how? Second, there is the problem of structure and agent. How far are France's failures to be traced to inherent characteristics of the French state and society, of France as a power and a military system, and how far are they to be attributed to problems of political and military leadership, whether a Ludovician failure to measure up to the standards of a Frederick the Great, or a Napoleonic hubris? Third, how are domestic aspects to be related to the warfare and international relations of the period? Was there a self-righting system within Europe that prevented any one power from becoming too strong? Were there other characteristics of the warfare and international relations of the period that harmed France? More generally, was this true of European developments of the period, for example the impact of the economic development of Britain, Austria and Russia, or France's declining demographic weight within Europe?

Although they do not have to be, structural interpretations are apt to be deterministic. Such an approach, however, is inappropriate for the whirligig of international relations and war of this period, and even the contemporary mechanistic notions – of balance, of power, of natural interests, and national characteristics in war – were tempered by an emphasis on the role of individual preferences and skills of particular rulers. More generally, it was unclear to contemporaries whether these notions were descriptive or normative.[17]

Concern about systemic and structural accounts returns attention to the specific and the individual. It offers the opportunity to join policies and military systems to circumstances, not with the former dictated by the latter, but in an interactive and dynamic relationship in which policies were chosen and systems developed, and debates over policy and system had weight and meaning. An unwillingness to adopt deterministic viewpoints should not mean that there are no attempts to advance conclusions or to discern broad trends, although it is, of course, necessary to be cautious not to reify the latter and give them explanatory weight, confusing description with explanation.

By placing an emphasis on contingency, and querying structural factors, it becomes unclear how best to take the cultural dimensions of the question, as indeed of other questions. These dimensions have been emphasized in studies of the Anglo–French rivalry. The French attachment to a moral economy and monocultural kingdom has been seen as less successful than the British embrace of mercantile capitalism and doctrinal tolerance. This has been linked to specific points, especially the British ability to borrow at a lower rate of interest, and more generally, the greater role of naval, colonial and mercantile interests and issues in British policy.[18] The latter was assumed to be the case by French critics of their own government. It can be seen as important in successive conflicts, especially the Seven Years' War and the Revolutionary and Napoleonic Wars.

There is clearly some weight in this analysis, although less than is frequently assumed.[19] In the case of France, royal preferences were serious, because the emphasis within French government and politics was not on matters maritime, commercial or colonial. French failure in the struggle for oceanic power, and thus dominance of the maritime trade of the world, has been traced in part to the impact of aristocratic political culture and priorities over those of bourgeois origin. Yet before this is seen as the failure of an entire governing order, it is worth stressing that, at various times, powerful ministers, such as Maurepas and Sartine, were committed to naval power and colonial expansion; and there was nothing pre-ordained about French priorities and policies, in Europe or outside. Furthermore, a cultural approach is not very helpful when considering French rivalry with Continental neighbours, particularly Austria, for there was no contrast comparable to that with Britain. This rivalry was central to French policy under Louis XIV and, arguably, until the close of the Seven Years' War, and again under the Revolutionaries and Napoleon. In addition, cultural concepts such as glory and honour were not uniform, or unchanging in their impact. It is important to ascertain what they meant to specific individuals or groups at particular moments. Such specificity subverts long-term models, and is difficult to incorporate into them.

A judgement of military capability and effectiveness has to take note of political priorities. It has been suggested that French policy became more focused on colonial issues from the 1740s,[20] but although there was more consistent attention to colonial questions, especially over Canada, and the fleet was built up, these were less weighty in French foreign policy in 1749–53 than the attempts to thwart Austrian, Russian and British schemes in Europe. In short, military capability in Europe was more important than its naval counterpart. The North American crisis with Britain in 1754 was a surprise to both powers, and it has been recently suggested that it was needlessly provoked by the aggression of French Marine officials.[21] The crisis led to a British attack in North America in 1755, but French diplomacy in 1755–6 was primarily devoted to a continuation of the earlier attempt to thwart Anglo-Austrian-

Russian schemes in Europe. Furthermore, Dupleix's expansionist policy in India was abandoned in 1754, and a provisional peace was reached with the British that winter.

The so-called Diplomatic Revolution of 1756 was, from the French point of view, as much, if not more, an opportunity to create a stronger position in Europe, taking forward long-held hopes of better relations with Austria, as a chance to improve the situation for waging war with Britain by ending the likelihood that Austria would support her. During the Seven Years' War, the French concentrated their military resources in Europe for more reasons than that of opportunity, and this is important in judging France's situation as a military power. In addition, the joint Franco-Spanish land attack on Portugal in 1762 appeared an effective way to strike at Britain through a vulnerable ally. French colonial policy changed only from the 1760s. Choiseul aggressively asserted imperial control over the French colonies after the debacle of the Seven Years' War, while French patriotism was increasingly focused on the confrontation with Britain.[22]

It is thus necessary to avoid reading from Britain's earlier maritime and colonial priorities in order to establish their French counterparts, and the degree and causes of French failure. Indeed, even if attention is focused on colonial gains, it is worth noting that the British triumph in the Peace of Paris of 1763 was an example of one of two types of Anglo-French peace treaty. In 1748 (Aix-la-Chapelle) and 1802 (Amiens), the British had to return colonial gains in order to compensate for Continental defeats, unlike in 1713 (Utrecht), 1763 (Paris), and 1814–15 (Vienna). In short, land victories could compensate for naval and colonial defeats. There was no reason in 1754–6 to anticipate that it would not be possible for the French to repeat their attempts during the 1740s to make Hanover and the Austrian Netherlands (Belgium) hostages for the fate of the colonies. The advance of French forces towards Hanover in 1741 had brought short-term gains, while Marshal Saxe's overrunning of the Austrian Netherlands in 1745–7 and his invasion of the United Provinces in 1748 had been very important to the peace settlement of 1748. However, the French alliance with Austria in 1756 closed the latter option, even while it eased France's strategic position in Europe: it was not necessary anymore for France (or Britain) to fight in Italy, on the Rhine, or on the north-eastern frontier. Strategy was therefore set by policy.

The effects of the Austrian alliance focus the issue of choice. Far from interests appearing clear-cut and alliances predictable, French policy, and that of the other powers, was shot through with unpredictability. Having decided not to fight Austria in 1727 and 1730, France attacked it in 1733 and 1741, before allying in 1756. More generally, France in 1700–60 fought Austria, Britain, the Dutch, Prussia, Sardinia and Spain. Its forces had to be prepared for a variety of tasks, far more so than those of Britain in Europe. France also had important alliances with each of these powers for at least part of the period.

Focusing on alliances, and on the manner in which they were negotiated (with no sense of structural necessity), directs attention to the role of choice, both by France and by other powers. France's conflicts and peacetime position can be understood, at least in part, in terms of the strength, determination and dynamics of alliances. This then invites the question whether failure should be best understood not as stemming from military conduct, or resources, or structural domestic characteristics, but rather in terms of these alliances, and the resulting enmities – an approach that can also be employed, for example, with reference to Germany and the two World Wars. Did France fail because of its alliance choices? Specifically, in general in the eighteenth century, France preferred to look to Poland, Turkey and Sweden rather than Russia; Bavaria (and, after 1756, Austria) rather than Prussia, and Spain rather than Britain.

Choices, of course, were linked to conceptions – by both France and other powers – of France's natural policy and conduct. These involved more than what in the twentieth century were termed 'realist' issues. Aside, for example, from 'prudential' considerations about Russia's limitation as a reliable ally, there were also notions of traditional obligations towards the Poles. Such notions were focused by the expansion of Austrian and Russian resources, power and pretensions in 1683–1721, an expansion that established Austrian power in Hungary, Italy and the Low Countries, and Russia on the Baltic. In combination, the military successes of the two powers at the expense of Turkey and Sweden changed European politics, ensured that Louis XV (1715–74) and Louis XVI (1774–93) had to deal with far more difficult problems in Eastern Europe than did Louis XIV (1643–1715), and arguably meant that France was likely to fail unless it could reach an acceptable agreement with the two states.

Much of the history of France as a European power over the following century can be seen in terms of an attempt to do so, first by resisting the pretensions of Austria and Russia, the major theme in 1688–1755, then by co-operating with Austria, and seeking to control it, until 1791, and next, in 1792–1809, by fighting Austria and Russia, in order to force them to adjust to a new settlement of Europe. French reluctance to accept the real or apparent pretensions of other major powers did not mean that French governments necessarily sought hegemony or dominance, but rather that it was difficult for the French to conceptualize a satisfactory stable response towards other states that were seen as inherently expansionist. The British faced a similar problem in Europe, but in 1793–1815 came to accept Austria, Prussian and Russian expansionism as the price for containing France. As a consequence, they could jointly invade France in 1814, and Blücher's Prussians could come to the aid of Wellington. Similarly, in the Second World War, the USA and Britain accepted Soviet expansionism, and again Poland was the victim.

In part, this difficulty of conceptualizing a satisfactory response to expanding states was a matter of French cultural politics: the definition of an international order and set of French goals within which it was possible to pursue allies and reach agreements. In particular, the French found it very difficult to accept that Russia was a major power and would pursue its own interests. The role of such assumptions was, and is, very important in creating warlike situations and in formulating wartime goals.

This argument cannot be pushed too far. Concern about French intentions and methods did not necessarily prevent co-operation with France, and could, indeed, encourage it. This was particularly seen under both Louis XIV, as in 1670–3, and Napoleon, as in 1810–11. It could be a matter of shared benefit or fear, and the two could also be reconciled in a willingness to accept the hegemonic pretensions of France. This was also to be true with Nazi Germany.

Alliance possibilities, alliances, and wartime co-operation were constricted, as much by opinion as by interest; and the latter cannot be understood separately from the former. It was particularly difficult to ensure a positive response on the part of other powers, if policy apparently stood for change. This was not only true of France. It took a century for Russia to force the Poles and Swedes to accept its hegemony, and this enforced compliance was important to the consolidation of military success. Acceptance of French dominance of Alsace, Lorraine, France-Comté, Artois and French Flanders also took many decades. Yet by 1748 this had been achieved, both because the French had been successful in war in 1733–5 and 1745–8, and because of a change in French aggrandisement: further gains in the Low Countries were not pursued after the War of the Spanish Succession (1701–14), and it was clear that the acquisition of the reversion to Lorraine after the War of the Polish Succession (1733–5) would not serve as the basis for additional gains to the east. There was a change of ethos and method in French policy. In 1748, Louis XV debarred his negotiators from holding onto conquests in the Austrian Netherlands, in effect confirming that France was content with its frontiers. France's borders were stabilized in a series of frontier treaties.

Partly as a consequence, there was a major shift between the reign of Louis XIV and the period *c.* 1750–*c.* 1790. The latter is generally seen as a period of failure in French policy and of military weakness, and indeed, it was then that France's traditional alliance system in Eastern Europe collapsed, and then that France's prestige as an international and military power was tarnished domestically. Yet there was also an important degree of success, one that highlights the conflicting analyses of success that were, and are, possible. France's alliance politics were more successful between 1756 and

1786 than previously and subsequently. It was no longer isolated, and its alliance system widened to encompass much of the European world. This is underrated because Britain was not decisively beaten, but the absence of any naval ally for Britain between 1748 and 1787 and its loss of America are, in part, instructive comments on French success. In addition, the Western Question had been settled, with the Bourbons dominant in Western Europe, and France clearly the major power. Neither had been true in the sixteenth and seventeenth centuries. This was more important than the size or success of the army. With the Western Question settled, French foreign policy tended to react to developments elsewhere, rather than direct them.

From 1756, France was linked to the crucial alignment in Eastern Europe: that of Austria and Russia. France benefited from a transformation in Austrian policy; Prussia, not France, was Austria's principal opponent in 1744–5 and from 1748. It was worth losing the battle of Rossbach to Prussia in 1757 as the price of alliance with Austria – a perspective on military history that is worth probing.

France's position in 1749–86 throws light on its policy both under Louis XIV and from 1792 until 1815. It was not that alliances were impossible then. England joined France against the Dutch in 1672 and in negotiating the Partition Treaties of 1698 and 1700. The Revolutionaries settled their differences with Prussia and Spain in 1795. Napoleon signed the Tilsit agreement with Alexander I (1807) and was able to reach an understanding with Austria. Yet these alliances had to be fought for, and most were very short-term – a factor that, in turn, affected France's other alliances, as well as attitudes towards France, and French foreign policy as a whole.

The contrast between the situation in the last decades of the *ancien régime* and that both earlier and later indicates that there was no inherent reason for the failure, but essentially, France failed as a great power because it was unable to operate alliance politics successfully. This was the cause, not the consequence, of its military failure. This failure in alliance politics was important both for its Continental position, and also in its maritime struggle with Britain. In the latter case, repeated failures in relations with Spain and the

Dutch fatally weakened the French, while the inability to sustain alliance with the Americans was also important. In 1779, the Spanish fleet had joined the French in attempting an invasion of England. In contrast, Napoleon's exploitative and insensitive treatment of Spain lessened and, in 1808, eventually destroyed its value as an ally, and thus gravely compromised France's opportunities as a global power. Britain was also given an opportunity to contest French power on land.[23] This was an example of the diplomatic exchange of the Napoleonic period which could have produced consequences other than those which did transpire.

Napoleon's provocative attitude towards his allies was the principal reason for his military failure, although French capability was also affected by improvement in the fighting quality and methods and military organization of its opponents. Having championed people's war in the 1790s, the French, in turn, were hit by it in 1808–13, particularly in Spain. The French envoy in Munich complained in 1809 in tones that would not have been out of place for France's opponents in the 1790s:

> Ces Autrichiens, ressemblent beaucoup plus à une horde de sauvages qu'à des soldats . . . Les Autrichiens se conduisent partout comme des brigands. Partout ils s'annoncent comme des libérateurs, ils forment des Comités d'insurrection et en même temps ils demandent des contributions énormes.[24]

More specifically, other powers learned to emulate or counter French organizational, operational and tactical methods. Indeed, it has been argued that a new 'era' in conflict began with the Franco–Austrian War of 1809, as other powers matched Napoleon's modernization of warfare, and even that that war was 'the first modern war'.[25]

The specific point can be contested, and indeed, it has been claimed that the campaigns of 1813 are a more appropriate starting point for modern war, but the more general assumption is valid. Far from seeing the initial warfare of 1792–4 as revolutionary and the remainder of the period 1792–1815 as a working through of themes, it is instead necessary to note the tempo of change after 1794, and, as with the First World War (see p. 217), the importance of developments far later in the period.

A focus on his opponents[26] offers an opportunity to re-evaluate Napoleon. So does an assessment of French generalship in the 1790s. It has, for example, been suggested that Napoleon was no greater in the strategic and operational arts than Dumouriez, nor a greater master of the central direction of war than Carnot.[27] The critical re-evaluation of Napoleon as a military leader, also seen, for example, in Owen Connelly's *Blundering to Glory: Napoleon's Military Campaigns* (2nd edn, Wilmington, DE, 1999), permits us to add to the analysis that Napoleon was a failure because of the character of his foreign policy the suggestion that it is necessary to offer a military account of his failure.

There is an obvious parallel with recent insights stressing the deficiencies of another paradigm of military success: the German army of the early twentieth century and its strategy, operational methods and relative fighting quality (see pp. 214–15). For Napoleon and, later, the Germans, effectiveness rested not on some devastating new military characteristic, bar high-tempo attack, but on the particularities of individual battles and campaigns, and on the relationship between the military and diplomatic strategies of the combatants.

At the same time, it is important not to take criticism of Napoleon too far.[28] His search for fame had an important military consequence, in that it contributed to the morale of his troops, and thus his ability to impose his will upon war. Napoleon's command of the respect and affection of officers and troops was a major resource. His ability to develop and use the corps in order to pursue the operational level of war helped Napoleon to wage effective campaigns.[29] To suggest that he was an excellent general is not to denigrate his predecessors and contemporaries in the 1790s, nor to ignore his faults and limitations as a commander, nor his serious deficiencies as a ruler. Equally, it is mistaken to offer an overly static account of Napoleon's generalship. The variety of his opponents and the pace of change was such that it was necessary to be continuously flexible and adaptable.

Napoleon has been faulted on this head, and his handling of the 1815 campaign was indeed maladroit: his failure to drive his opponents apart in order to defeat them in detail, and his lack of flexibility on the battlefield at Waterloo. However, it is important to note earlier limitations in the French military system. Had the factors of weather and distance been comparable to those in Russia in 1812, then it is possible that the failure of the Napoleonic system would have been dated earlier. Such factors certainly hit the French campaign in Poland in 1807. Napoleon's effectiveness within a certain range was effectively challenged in 1813. That year, he suffered from the consequences of the losses in 1812. In particular, Napoleon's loss of cavalry in 1812 meant that he could not exploit his victories at Lützen and Bautzen in 1813. That year, Napoleon also suffered from the limitations of his marshals as independent commanders, and this was exploited by his opponents in the Trachenburg strategy. Yet such an analysis should not disguise Napoleon's failures. Dresden was not a sweeping victory, while Leipzig was a disaster. His opponents fielded effective forces, and Napoleon in 1813 could not repeat the Revolutionaries' achievement in 1794.

The number of Napoleon's opponents was obviously important in 1812, 1813 and 1814. That was a product of political failure. He suffered from his unwillingness to compromise and thus to bring any stability to the inherently unstable French hegemony across much of Europe. This instability was exacerbated by the invasion of Spain, and the situation collapsed as a consequence of failure in 1812–13. The large numbers of Germans, Italians and Poles that invaded Russia in 1812, and the Saxons and Danes who supported him in 1813, were led to defeat. This defeat exposed Napoleon's character flaws. To be unwilling to compromise when triumphant was serious enough. To fail to do so when defeated was far more so. In January 1813, for example, Napoleon harshly rejected Prussian terms. That summer, he mishandled Austria, and in early 1814 he ruined the chance to use the negotiations at Châtillon to bring peace.[30]

188

This would have mattered less had there been an issue both to divide his opponents and create a different target, as Poland had done in 1793–5. The equivalent for Britain might have been failure in the war of 1812–15 with America, American conquest of Canada, and the threat of American intervention in the West Indies or Ireland; but none of these were plausible. There was no equivalent to Poland, Turkey had ended its war with Russia in 1812, and the Americans lacked long-range amphibious capability or the unity and resources for a major war.

Napoleon was an effective general, but the capability gap between French forces and their opponents that had been so important in the late 1790s and early 1800s had even then not been unbridgeable, as a number of commanders had shown, most obviously the Russian Count Suvorov in Italy in 1799. The French lacked a lead comparable to that enjoyed (although not without anxiety) by the British at sea after Trafalgar, and the circumstances of land warfare were not such as to permit a replication of a lead of this type. The combination of this military precariousness with political opposition to French hegemony required skilful management, both militarily and politically. His apologists notwithstanding, Napoleon could not provide this.

There were therefore choices in French foreign policy throughout the period 1688–1815, and these greatly affected military power, prefiguring the consequences of breakdowns in German–Soviet and Soviet–Chinese relations during the Second World War and the Cold War respectively. Could Napoleon have maintained the Tilsit agreement, successfully incorporated Iberia, peaceably or by conquest, into his system, and then forced Britain to terms that left France a major role outside Europe? Would his troops have been more successful had they invaded England in 1805 rather than Russia in 1812? More generally, Napoleon lost sight of the politics that war was supposed to further.

Different scenarios should not be dismissed with reference to structural limitations in French finances or other similar factors. France was different to Britain as an economy, a society and a state. This affected, but did not determine, policy choices, resource availability and international responses. Strategic options remained, as both the British and the French showed in the War of American Independence.[31] Furthermore, as already suggested, there may, after all, have been no need to choose between European and global roles. A Jacobite Britain might have accepted, or had to accept, a North Atlantic world dominated by France.

Focusing on diplomatic and military failures makes sense of the contrast between long-term 'structural' characteristics and the very varied fate of France as a great power. France's failure was as much the consequence of political, diplomatic and military contingencies and decisions, as of structural or socio-economic factors. Thus, the RMA of the 1790s could not overcome for long the political consequences of French policy. Furthermore, the West was left by 1815 without a hegemonic power, and also with a political settlement that left much of Europe stable, in so far as territorial boundaries were concerned, and with both Britain and Russia free to pursue expansionist interests elsewhere in the world. This resolution of the struggle in the Western 'core' was very important to the subsequent century of European imperial expansion.

NOTES

1. For the theoretical perspective, see P.E. Tetlock and A. Belkin (eds), *Counterfactual Thought Experiments in World Politics: Logical, Methodological and Psychological Perspectives* (Princeton, NJ, 1996). For a theoretical introduction and a series of case studies, see N. Ferguson (ed.) *Virtual History: Alternatives and Counterfactuals* (1997). See also E. Durschmied, *The Hinge Factor* (1999).
2. J.M. Black, *Culloden and the '45* (Stroud, 2000).
3. J.P.C. Laband and P.S. Thompson, *Field Guide to the War in Zululand and the Defence of Natal, 1879* (Pietermaritzburg, 1983); I. Knight, *The Anatomy of the Zulu Army: From Shaka to Cetshwayo, 1818–1879* (1995).
4. E. Gibbon, *The History of the Decline and Fall of the Roman Empire* (ed. J.B. Bury, 7 vols, 1896–1900), IV, 166–7.
5. T. Bartlett and K. Jeffery (eds), *A Military History of Ireland* (Cambridge, 1996).
6. S. Conway, *The War of American Independence 1775–1783* (1995).
7. For a similar approach to the politics, see J.C.D. Clark, *The Language of Liberty 1660–1832: Political Discourse and Social Dynamics in the Anglo-American World* (Cambridge, 1993).

190

8. E. Gibbon, *Decline and Fall*, IV, 166–7, VI, 98.

9. J.M. Black, *War for America: The Fight for Independence, 1775–83* (Stroud, 1991); P. Mackesy, *The War for America 1775–1783* (2nd edn, Lincoln, NB, 1993).

10. J.M. Black, *Pitt the Elder* (2nd edn, Stroud, 1999).

11. R. Kohn, *Eagle and Sword: The Beginnings of the Military Establishment in America* (New York, 1975); E. Skeen, *Citizen Soldiers in the War of 1812* (Lexington, KY, 1999).

12. For a stress on the plasticity of America's frontiers and identity, see D.W. Meinig, *The Shaping of America: A Geographical Perspective on 500 Years of History* (3 vols to date, New Haven, CT, 1986–98).

13. G.A. Smith, 'Storm Over the Gulf: America's Destiny Becoming Manifest', in *The Consortium on Revolutionary Europe: Selected Papers, 1994* (Tallahassee, FL, 1994), pp. 510–16.

14. See, for example, J. Gill, *With Eagles to Glory: Napoleon and His German Allies in the 1809 Campaign* (1992); F. Schneid, *Soldiers of Napoleon's Kingdom of Italy: Army, State, and Society 1800–1815* (Boulder, CO, 1995).

15. J.M. Black, *From Louis XIV to Napoleon: The Fate of a Great Power* (1999).

16. P. Wetzler, *War and Subsistence: The Sambre and Meuse Army of the French Revolution* (Cambridge, 1988); J.A. Lynn, *The Bayonets of the Republic: Motivation and Tactics in the Army of Revolutionary France, 1791–94* (2nd edn, Boulder, CO, 1996); P.H. Wilson, *German Armies: War and German politics 1648–1806* (1998), pp. 338–9.

17. J.M. Black, *The Rise of the European Powers 1679–1793* (1990).

18. See, for example, J. Brewer, *The Sinews of Power, War, Money, and the English State, 1688–1783* (1989), H. Bowen, *War and British Society 1688–1815* (Cambridge, 1998), and F. Crouzet and P. Butel, 'Empire and Economic Growth: The Case of Eighteenth-century France', *Revista de Historia Económica*, 1 (1998), pp. 29–92.

19. J.M. Black, *Britain as a Military Power, 1688–1815* (1999). On foreign policy, see J.M. Black, *A System of Ambition? British Foreign Policy 1660–1793* (2nd edn, Stroud, 2000), and *British Foreign Policy in an Age of Revolutions, 1783–1793* (Cambridge, 1994).

20. D.A. Baugh, 'Withdrawing from Europe: Anglo-French Maritime Geopolitics', *International History Review*, 20 (1998), pp. 1–32.

21. F.W. Brecher, *Losing a Continent: France's North American Policy, 1753–1763* (Westport, CT, 1998).

22. E. Dziembowski, *Un nouveau patriotisme français, 1750–1779: La France face à la puissance anglaise à l'époque de la guerre de Sept Ans* (Oxford, 1998).

23. C. Hall, *British Strategy in the Napoleonic War, 1803–1815* (Manchester, 1992); R. Muir, *Britain and the Defeat of Napoleon 1807–1815* (New Haven, CT, 1996).

24. L.-G. Otto, 'L'Ambassadeur Otto de Mosloy d'après des lettres inédites', *Revue d'Histoire Diplomatique*, 69–70 (1955–6), pp. 18, 21.
25. R.M. Epstein, *Napoleon's Last Victory and the Emergence of Modern War* (Lawrence, KS, 1994), pp. 2–9, 171–83.
26. G. Rothenberg, *Napoleon's Great Adversaries: The Archduke Charles and the Austrian Army 1792–1814* (Bloomington, IN, 1982).
27. P. Griffith, *The Art of War of Revolutionary France 1789–1802* (1998), p. 11.
28. For a positive recent assessment, see D. Gates, *The Napoleonic Wars 1803–1815* (1997).
29. J. Luvaas (ed.), *Napoleon on the Art of War* (New York, 1999), p. x.
30. A masterly criticism of Napoleon is offered by P.W. Schroeder, *The Transformation of European Politics* (Oxford, 1994), e.g. p. 343. For a defence, claiming that Napoleon was driven by events and in large part the victim of aggression, R.M. Epstein, 'Revisiting Napoleon: For and Against', *Consortium on Revolutionary Europe 1750–1850: Proceedings 1998* (Tallahassee, 1998), pp. 269–81.
31. J. Dull, *The French Navy and American Independence: A Study of Arms and Diplomacy, 1774–1787* (Princeton, NJ, 1975); N.A.M. Rodger, *The Insatiable Earl: A Life of John Montagu, 4th Earl of Sandwich* (1993).

EIGHT

Technology and Western Dominance
1815–1980

Periodization is always problematic, and military history is no exception. This is true not only of the attempt of some of those interested in present and future warfare to deny the relevance of past circumstances, but also of the understanding of the latter. Periodization entails the creation of categories. It directs attention to differences between them, and then to the process by which these differences were made. As already suggested in Chapter Four, the notion of the early-modern period is not without numerous difficulties. This is also true of what follows: the late-modern period, an age that can be variously defined and dated.

Generally, the late-modern in military terms is sub-divided, so that, as today is approached, the number of chronological categories increases and their length diminishes. The twentieth century is particularly thoroughly dissected. The implication, sometimes made explicit, is clear: the rate of change was greater in more recent times, and this requires both more interpretation and more categories. Yet this approach may be queried, and we can ask whether it would not be profitable to examine the late-modern period as a unit similar in length to that of earlier chronological categories. This is the method that will be adopted in this chapter. It reflects an attempt to assess continuities and to move beyond the undoubted role of technology in altering the manner of conducting war. The last has been probed in many excellent works, and there is no need to reprise the importance of the military applications of a technological-industrial-organizational nexus that was inherent to the process of Western change in this period.[1]

THE GLOBAL DIMENSION

Instead, adopting the theorem of war as the pursuit of politics by other means, and considering politics in terms of both international relations and internal control, it is suggested that the late-modern period can be dated as ending in the 1980s. That decade saw the completion of a RAM (Revolution in Attitudes towards the Military) that made expansionist war by the major powers (but not all states) seem no longer acceptable. The beginning of this period is less clear, but a measure of coherence can be offered for the late-modern period by suggesting that this was the age when Western powers and Russia/the Soviet Union dominated expansionism and were generally able to prevail, albeit not without checks. Some checks, such as the Chinese repulse of British and French warships at Dagu in 1859, and the Zulu victory over the British at Islandwana in 1879, were rapidly reversed, but others, such as the disastrous Italian defeat by the Ethiopians at Adua (Adowa) in 1896, had a longer impact.

Even successes could indicate a strength of resistance that discouraged further Western activity. The Boxer rebellion in China in 1900 revealed the strength of national sentiment, or at least xenophobia, and can be seen as crucial in ending the carve-up of China between the great powers that had been advanced by the allocation of spheres of influence in 1897–8. Although the Western positions in both Tientsin and Beijing were relieved, the challenge to Western power had been spectacular, and the campaign had not been easy: the first attempt to relieve Beijing was unsuccessful, and the garrison in Tientsin had been hard pressed.[2]

Nevertheless, defeats and checks did not lead to a widespread overthrow of the Western position. Instead, their consequences were sealed off. For example, British defeat in the First Afghan War of 1838–42, specifically the failure to establish garrison rule on the

194

Indian model, was followed by the diplomatic isolation of Afghanistan and the strengthening of the Indian border, not by an Afghan invasion of India comparable to that in the 1750s.

Despite checks, the contrast with the seventeenth and eighteenth centuries was readily apparent. The process of Western expansion was both more successful and more continuous. Furthermore, this was true of a variety of military environments, themselves composed of the interaction of ecological, political, social and more narrowly military circumstances. The situation in Africa and East Asia was particularly different to that in the eighteenth century, although in South Asia and Australasia the shift towards European expansion occurred earlier than 1800.

There was a mutually reinforcing process of causality in Western expansion. As victory and conquest became easier, so expansionism and a sense of superiority were encouraged. There was also greater interest in imperial conquest. Campaigns were followed by the public back home, and military interest in the nature of colonial warfare rose. Col C.E. Callwell's *Small Wars: Their Principles and Practice*, published in 1896, appeared in its third edition in 1906. Based largely on British campaigns, the book gave details of how to operate against irregulars.[3] Conversely, a sense of mission helped in directing and sustaining resources that made victory easier.

The comparative context is important. Eighteenth-century China had also been interested in expansionism, had a sense of superiority, and was militarily successful, indeed gaining vast tracts of territory. However, this was lost in the nineteenth century, such that in 1857 an Anglo-French force seized Canton, and in 1860 Anglo-French forces occupied Beijing after defeating Chinese forces. Two years earlier, most of the northern bank of the Amur River had been ceded to Russia in the Treaty of Aigun. Sino–French hostilities in 1884–5 left the French dominant in Vietnam and exposed anew the vulnerability of the Chinese coasts.

The reasons for Chinese military decline are unclear. It can be argued that eighteenth-century success was a cause of later failure. The *generally* peaceful nature of China in 1770–1835 contrasted with the frequency of conflict and pace of military change in Europe

195

and India. Using the challenge/response model, it can be suggested that continued Zungar pressure would have kept the military centrestage and maintained the viability of the Chinese armed forces. The argument has to be handled with care. Had the Zungars remained powerful they might have successfully contested the Chinese position in Mongolia and Tibet, possibly as well as Chinese control of Gansu and even Shaanxi. This would have weakened China prior to the onset of European pressure, and might also have made it far harder to suppress rebellions. Campaigns against Tongking (1788) and Nepal (1792) were small-scale, certainly in comparison to the warfare with the Zungars. The major conflict of the period was due to the huge and very costly White Lotus rebellion of 1796–1805 in Shaanxi. Unfortunately, there is no good monograph on the subject, but it can be seen as a turning point in Manchu history from which the regime never really recovered.[4]

A sense of a relative shift in capability was seen in many parts of the world with the borrowing of Western military technology and, albeit to a lesser extent and with greater difficulty, organization. This was not a novel feature. Indeed, Western cannon had been in demand in the Middle and Near East from the fifteenth century. Hungarian and German gunners helped Mehmed II at Constantinople in 1453, and in 1472, Uzun Hasan of Persia requested cannon, gunners and arquebuses from Venice. Yet this process had been limited in the sixteenth and seventeenth centuries for a number of reasons. First, there had been alternative sources of advanced weaponry and military skills, particularly Ottoman Turkey in the Islamic world, whether for Aceh in Sumatra or Bornu in the African Sahel. Second, the apparent need for such technology was limited and, more particularly, borrowing was largely restricted to specialist weapons and units. This was true, for example, of the Chinese use of Western cannon in the seventeenth century,[5] and of the Ottoman creation of Western-style units in the eighteenth.[6] Thus, borrowing from the West contributed to the dominant style of composite forces, rather than dominating, let alone displacing them.

This situation altered in the nineteenth century, as states created substantial Westernized forces or Westernized aspects of their entire military.[7] The potency of Western military methods was demonstrated not only in clashes with Western powers, but also in the use of Western-style forces and equipment in conflicts elsewhere in the world. Thus, Western-officered units were deployed against the Taipings, while British steamships were used to transport Chinese forces during the same conflict.[8]

Japan introduced conscription in 1873, and followed it up by employing German talent to help train the army leadership. The Imperial Japanese Navy, established in 1868, instead looked to the British. In 1882, when the British attacked Alexandria, they fought Egyptians equipped and deployed in a very different fashion to those who had been beaten by Napoleon at the Battle of the Pyramids in 1798.

This Westernization, however, was incomplete in a number of respects. In many states, such as China and Persia, it was small-scale until late. Westernization also very frequently affected weaponry rather than the organization, command structure and ethos of armies. Westernization overlay political and social structures and attitudes different to those that were dominant in the West.[9]

The development of China's modern military industry in the nineteenth century is an excellent example of the problems which arise when importing technology and using other means of overcoming a technological deficit. The first attempts to start military industries using technology from the West were the Self-Strengthening enterprises which began in about 1860. The most general problem with this policy – at least in the eyes of some – was the attempt to preserve Chinese culture and only bolt on Western technology, although anti-foreigner nationalism greatly restricted options. Self-Strengthening began after two defeats by the West and a disastrous series of internal rebellions. It was never a national policy – the initiative was taken by leading provincial officials, who adopted a form of state entrepreneurship sometimes known as 'official supervision, merchant management'.

More specific problems arose in particular enterprises. At the Jiangnan arsenal, founded in 1865, Remington rifles were manufactured, but they were of poorer quality and more expensive than rifles obtainable in the West; steam-powered gunboats were launched there from 1868.[10] At the Fuzhou Naval Dockyard, founded in 1866, the ships launched were obsolescent when they left the slipway, although their quality improved: steam-powered but fully-rigged wooden ships were superseded by more advanced vessels. Attempts were made to close the technological gap by buying weapons and ships, and by employing Western experts, but with only limited success. Nevertheless, although the Self-Strengthening programme is often accounted a failure – reference is made to Japan's experience and to China's defeat in the Sino–Japanese War of 1894–5 – too little attention is given to the enormous strides made by China in this period.[11]

Later, attempts to improve the Chinese military on a Western pattern, those of Yuan Shikai and Chiang Kaishek in the early twentieth century, suffered from the absence of a political unity necessary in order to man and fund a national army. Both tried to use the military to give unity to the state, but were confronted by powerful divisive and fissiparous tendencies. The creation, for example, of an army ministry in 1906 was helpful, but not enough. Military modernization played an important role in bringing down the Qing. In 1911, New Army officers in Wuchang began the revolution,[12] and Yuan Shikai, the creator of the effective Western-style Beiyang Army, was instrumental in the success of the revolution. In 1912, he became President of the new republic.[13]

Different attitudes between Western and non-Western societies did not necessarily preclude victories for Westernized forces. This was most obvious with the Japanese victory over the Russian Baltic Fleet at Tsushima (1905), the sole example of a decisive clash between modern battleships in the twentieth century. Japanese success in the 1904–5 conflict had been prefigured by the defeats of European armies by opponents employing state-of-the-art European weaponry, most obviously the wars of liberation against Britain and Spain in the New World (between 1775 and 1826), and Boer successes against the British in South Africa in 1881 and 1899. Japanese victory created more of an impact because they were a non-Western people, and had not been seen as a major military force (Western observers expected Japan to fail against China in 1894–5), and because Japan became a leading naval power and expanded territorially. This was also true of the USA at the cusp of the nineteenth and twentieth centuries, but the Americans were a Western people. In addition, international attitudes to the USA and other independent states of the New World had been formed earlier during the nineteenth century, when they had not been significant naval powers and had not sought trans-oceanic expansion.

Accepting the complexities and limitations of the concept, it is nevertheless the case that Westernization can be used to provide a basic identity to the period 1815–1980 in terms of its military history. This concept is of additional value, because much of the warfare of the period involved the expansion of Western empires and, subsequently, decolonization struggles. Compared to the early-modern period, the relative importance of West v. Rest conflicts, as opposed to West v. West and Rest v. Rest, rose, although West v. West wars were very important, and Rest v. Rest conflicts remained more prominent than is frequently appreciated, even during the heyday of Western imperialism.

Treating the nineteenth and much of the twentieth century as a unit is not intended to underrate differences and shifts during this period. These related to the purposes of conflict, the nature of military organization, the tactics used, and the weaponry employed. Elsewhere, the order in which these are generally discussed suggests a possibly misleading prioritization. By putting weaponry last in my list, it is suggested that it was instrumental in conflict, rather than necessarily being central to it. This may appear a bizarre assessment of an age that began with muskets and ended with hydrogen bombs and inter-continental missiles, but it is necessary not to put the means of conflict first. The essential problems of how best to employ force to control people and territory did not alter, even when the capacity existed to destroy the world. This can be seen empirically – the possession of nuclear bombs did not bring the Americans victory in Vietnam or the British at Suez (1956) and in Aden and Ulster – and also by considering the extent to which weaponry does not offer short cuts to the problems of asymmetrical

199

warfare or the issues of strategic choice, political determination and alliance politics. Instead, if anything, irrespective of these limitations, weaponry became a greater problem for states, especially in the twentieth century. This was not so much due to the creation, with nuclear weaponry, of a category of weaponry that made conflict too risky, ensuring that it was impossible during the Cold War to obtain a military verdict, other than to specific struggles in particular hot spots. Rather, the cost of weaponry and of military systems in wartime became such that it was not possible for most states to maintain state-of-the-art forces, still less to sustain war.

As a consequence, in order to wage war, it was necessary to mobilize resources and direct society to an extent that undermined economies and social and political systems. This was especially true of the First World War, which gravely weakened European liberal democracies. The impact of war and the destructive capacity and cost of weaponry did not, however, lead to an end to conflict. Within a quarter-century of the close of the First World War, a major war broke out again in Europe, while, after the Second World War, the USA employed its forces in conflict to a considerable extent during the period 1950–72, albeit without using the full range of its military power.

The high rate of military preparedness and the readiness to resort to war, both despite their cost, in the nineteenth and early twentieth centuries directs attention to the purposes of preparation and conflict. The partial Westernization of the military around the world was accompanied by an extension not only of Western economic processes, but also of Western governmental forms and political ideologies. This process was aided by the variety of 'Wests' that were on offer.

In the nineteenth century, the incorporating model of Western imperialism – shared economic interests and a degree of citizenship – was of interest to some non-Western elites, or under the pressure of differences in military capability, or indeed even defeat by Western powers, could be made so. Thus, the British were able to win some support in India, Malaysia and Nigeria, the Russians in parts of

200

Central Asia, and so on, although it is important not to underrate the role of force in gaining support.[14]

Irrespective of this political support from elites, colonial powers also benefited greatly from the recruitment of local troops, spectacularly so with the French conquest of most of West Africa in the 1880s and 1890s. About half of the British force that attacked the Zulus in 1879 were Africans in the Natal Native Contingent. The officers were white colonists. The African horsemen were especially useful.[15] The use of native troops was also crucial to the British position in South Asia. Once the Punjab had been conquered in the 1840s, its Sikh troops proved a significant accession of strength for the British. Indian troops were used in expeditions to a variety of destinations, including China and Ethiopia. They were useful, for example, in the conquest of north Borneo and Labuan in 1838–46.

When, in 1889, at the height of the 'Great Game' for predominance in Asia, Col Grombtchevsky met Capt Francis Younghusband at Khaian Aksai, the first boasted that his Cossack escort would not need supplies to cross the mountains into India, while the second tried to impress with a drill performance by his Gurkhas. Two years later, the British successfully invaded Hunza with 1,000 Gurkha and Kashmiri troops under sixteen British officers. The 'Great Game' would have been impossible without massive local support, although even then it contained a powerful element of fantasy. Sir Garnet Wolseley, the fireman of British imperial power, described plans to advance into Central Asia as 'the dream of a madman whose head is filled with military theories drawn from the time of Xerxes and Alexander the Great'.[16]

The collective ideology of the 'West' also offered nationalism, which could encourage and become the language of anti-imperial action, as well as progressive notions of religious toleration, liberalism and constitutionalism which could help create or strengthen nations resisting imperial control. In the twentieth century, this broadened out with the addition of a Communist creed with universalist aspirations.

Thus, political control was contested in terms of Western ideologies, although these overlay and interacted with other identities. The overall effect was to encourage an apparent standardization of state forms across the world. In the nineteenth century, polities seeking to resist Western imperialism, such as Siam, sought to acquire and show the characteristics that Western powers regarded as proof of statehood. This involved, for example, the assumption and internalization of Western notions of territoriality.[17] At a basic level, the map had to be filled. This was because the maps produced for Western governments and purchasers ignored or underrated native peoples and states, presenting Africa and other areas, such as Oceania, as open to appropriation. This mapping helped to legitimize imperial expansion, to make the world appear empty, or at least uncivilized, unless under European control.[18]

The assertion of statehood on the Western pattern had several military consequences. First, it encouraged Westernization. As with the creation of the national 'Continental Army' by the American revolutionaries,[19] as part of their process of state creation, so other polities created national standing forces armed and, in part, organized on the Western model.[20] This was true, for example, of the revival of Turkey prior to the First World War: nationalism, military modernization and popular mobilization combined.

Second, the assertion of statehood led to conflict, as states sought to monopolize power within their area of claim, to expand this area, and to create and clarify frontiers. This was true both of states that maintained their independence in the nineteenth century, such as Turkey, Afghanistan, Ethiopia, Japan and Siam,[21] and of those that lost it or had it limited, such as Nepal, the Sikh state in the Punjab, Kashmir, Egypt and the Sudan. Thus, for example, Japan annexed the vassal Ryukyu kingdom in 1879, renaming it Okinawa, while Abdur Rahman (1880–1901), the 'Iron Amir' who united Afghanistan, transformed the militias into a standing army and defeated a series of risings including by the Ghilzais and the Heratis in 1886, in Badakshan in 1888 and among the Hazara in 1891. He seized Asmar in 1892, while Kafiristan was forcibly converted to Islam in 1895.

China needed to defend rather than to clarify its frontiers. After the chaos of the years of the Taiping rebellion, there was a process of reaffirmation of control over border areas that had also rebelled. Muslim rebellions in south-west (1856) and north-west (1862) China, the Yunnan and Shaan-Gan uprisings, had been defeated by the close of 1873, while the rebellion in Chinese Turkestan that began in 1863 was defeated in 1876–7. As with the Chinese conquest of the region in the 1750s, this was a formidable organizational achievement. The logistical task was very difficult. The Chinese were helped by divisions among their opponents and by the death of the ablest leader, Ya'qub Beg, in 1877.[22] In 1881, China persuaded Russia to restore the Ili Valley, which it had seized in 1871. Chinese authority over Tibet was forcibly reimposed in 1910, only for the troops to be expelled the following year.

More generally, the process of assertion of statehood resumed with decolonization, albeit with the added complication of dealing with the legacy of Western imperial rule, including frontiers that did not match ethnic boundaries. The last, for example, was a major problem in east-central Africa, helping to cause civil conflict in Rwanda and Burundi in the 1990s, and to link these conflicts to instability in neighbouring Zaire. The suppression of Armenian and Kurdish aspirations was a cause of conflict in the Middle East. In contrast, in Latin America, where early nineteenth-century decolonization was followed by tension and, at times, conflict over boundaries, particularly with the War of the Pacific (1879–83) and the Chaco War (1932–5), ethnicity was not an important issue.

It would no more be appropriate to attribute all conflict in the nineteenth and twentieth centuries to state-formation on the Western model than to suggest that the Western creation of a global economic system was responsible for slavery. Both conflict and slavery long preceded the Western impact. However, the impact of state-formation encouraged conflict, particularly in certain areas. First, regions where individual imperial powers had held sway were divided between competing states that waged war. Just as this had been a problem after the fall of the western Roman empire, it also affected Latin America after the collapse of Spanish rule in the early nineteenth century, the Balkans after the fall of the Turkish and Austro-Hungarian empires (with, for example, the Hungarian–Romanian war in 1919), South Asia and Palestine after the end of the British rule in 1947–8, and the Caucasus after the collapse of the Soviet Union in the 1990s.

Second, the creation of permanent military forces as a crucial indicator of statehood led to the employment of these forces to maintain unity and prevent separatism. This was not only true of the Third World in the twentieth century, but also of the West in the nineteenth. The use of force to suppress separatism in the USA in 1861–5, and also to subdue both Garibaldi's private army and regional social protest in Italy, can be seen in this light, as, after the First World War, can the operations of the Red Army in the Ukraine, the Caucasus and Central Asia,[23] and the suppression of regional separatism in Germany. Conversely, the Mexican army failed to suppress Texan secessionism in 1836. It was more successful against regional insurrections in 1911 and 1915. These processes became far more apparent in the Third World after decolonization, because many of the new states – for example Burma, India, Indonesia, Iraq, Nigeria, Pakistan and Zaire – lacked any real cohesion, certainly as far as frontier regions were concerned.

As suggested in Chapters Two and Three, the political purpose of military forces helped to determine their organization. It also conditioned their tactics and strategy, for the identity of the opponent was constructed in political terms. In part, this was simply a matter of the obvious question of whom one fought. For example, Japan's decision to attack the USA and the European colonial powers in 1941, rather than to develop earlier clashes with the Soviet Union into a full-scale war, necessarily affected what the Second World War meant for the Japanese military. This ranged from the need for the army to prepare for conflict in forested terrain rather than areas better attuned for mechanized conflict, especially Manchuria, and the balance of investment between army and navy, to the potential for long-range expansionist warfare. To underline the role of contingency, it was also far from inevitable in 1894–1905 that Japan would fight China and Russia, but not Spain over the Philippines, or that Chiang Kiashek would concentrate on fighting the Communists in 1930–4, rather than on resisting Japanese advances.[24]

The role of the construction of the target in affecting tactics and strategy was not restricted to the choice of the opponent. Instead, as with the causes of war, there were very different attitudes, ranging from the treatment of the opponent as sub-human to a mutual respect that encouraged an emphasis on professionalism. The opponent as sub-human, in contrast, led not only to brutality, but also to a targeting of the civilian population. This was seen, for example, in Japanese conduct in China, leading to episodes such as the Nanjing Massacre of 1937. It was also the case with German warmaking against the USSR in 1941–5, which was, in *some* respects, a transfer to Europe of methods hitherto used largely in imperialist wars elsewhere, although many aspects of German brutality were novel. This was true not only of the scale and systematic character of the slaughter of foreign civilians, but also of the massacre of German civilians. Furthermore, as an additional indication of Nazi viciousness and willingness to employ violence without qualification, large numbers in the German armed forces were executed: 30,000 or more, compared with fewer than 200 in the First World War.

The net effect of Nazi policies was to ensure that the military had to face insurgency by civilian populations. This was not new in Europe. Napoleon's forces encountered the same problem in Calabria and Spain,[25] and, in the war of 1870–1, the invading Prussians were resisted by French *franc-tireurs*. Conflict against an entire society can, in part, be linked to the suppression of a continuation of pre-modern peasant rebellions, as with the Nian rebellion in China in 1852–68.[26] Increasingly, however, it was a major effect of the notion of total war which became more common in the twentieth century. This involved naval blockade and the aerial or rocket bombardment of civilian populations. British incarceration of Boer civilians in the Second Boer War (1899–1902) ensured that the treatment of civilians was an issue from the outset of the century.

Total war also increased the military burden of occupying territory. Thus, the relative importance of obtaining a decisive battlefield result was lessened. To a certain extent, an analogy with fortification and siegecraft can be drawn, in that the role of victory in battle has always been qualified by the need to take control of territory. However, in total war, such control became very different. First, the opponent of invading and occupying powers was no longer solely or largely regular forces, as it was with the garrisons of fortified positions. Instead, a portion of the civilian population was involved. This shift diminished the predictability of symmetrical conflict, involved in conflict with regular forces, and made the whole civilian population potential opponents, because it was very difficult to assess who, in fact, were so. Second, in place of the need to gain control of regular fortifications, there was a situation in which every village, farmstead and wood was potentially a fortress. In September 1999, when the Russians threatened the breakaway republic of Chechnya in the Caucasus, President Maskhadov called on all men to mobilize themselves, and declared 'every village must be turned into a fortress'.

The consequence was a diminution in the effectiveness of regular forces, especially if poorly trained for counter-insurgency operations, as was the case with most armies and air forces. It was not the case that guerrilla operations were necessarily successful. Indeed, many popular risings and guerrilla forces were suppressed, more than is generally appreciated. For example, in 1931–3, the Japanese defeated very numerous guerrilla forces in Manchuria, while in 1946–59 – a period that is sometimes seen as successful for such forces, with the Viet Minh defeating the French in North Vietnam and Batista's regime failing to crush Castro – there was success not only against Communist movements in Greece and the Philippines, but also against anti-Communist guerrillas in Albania, the Baltic republics, Bulgaria, the Ukraine and Yugoslavia, while postwar Republican attempts to challenge the Franco regime in Spain were defeated, as was a nationalist rising against the French in Madagascar in 1947, and a rising the same year against Chiang Kaishek by native Taiwanese.[27] Some of these conflicts involved substantial forces and heavy casualties, although many are obscure. This is true, for example, of Soviet campaigns in the Baltic States and the Ukraine in the late 1940s. It has been suggested that the Soviets lost 20,000 men in suppressing opposition in Lithuania alone.

Thus, popular warfare did not automatically win. It did, however, greatly complicate the task for governments and regular forces, ensuring that an understanding of conflict as conduct between such forces was increasingly inappropriate in the twentieth century.

1815–1914

This is a long way from a conventional account of warfare in the period 1800–1980, with a stress on major conflicts. It is fair to ask how wars such as the Napoleonic conflicts, the Crimean War, the Wars of German Unification (or Prussian Aggression), the First World War, and the Second World War can be fitted into such a context. Each, indeed, centred on the clash between regular forces and yet each also showed the primacy of politics in war. For example, a major difference between the Napoleonic wars and the two World Wars was that Napoleon was able to negotiate treaties with all of his major opponents, in the case of Austria no fewer than three times. The wars therefore had an episodic quality, it was possible to achieve political settlements, and the creation and sustaining of coalitions against France was important to the course of the conduct and to strategy. In the two World Wars, in contrast, there was far less wartime negotiation between combatants, and less possibility of persuading powers to change side. In the 1790s, the Spaniards and Dutch, once defeated, became allies of France against Britain, and in 1812 Napoleon invaded Russia with significant assistance from former opponents, such as Prussia.

In the World Wars, there was use of the resources of conquered areas, and there were hostilities between Vichy France and Britain, but there was not a creation of new alliances resting on governments changing sides after defeat. Indeed, the failure of the Germans and their allies to win the support of defeated powers – for example Belgium in 1914, Serbia in 1915, Romania in 1916, Russia in 1917, and the large number of states defeated in 1939–41 – was a serious weakness; the willingness of Vichy French troops to resist the British in Syria, Lebanon and Madagascar in 1941–2 suggests what could have been done. Similarly, the Germans won Croat co-operation in conquered Yugoslavia in 1941, and also created a collaborationist Serbian government under General Nedic, but failed to take due advantage of their talks with Draza Mihailovic's Serb Chetniks that year. Furthermore, again unlike the Napoleonic wars, conflict was continual in the two World Wars. There were neither truces between the major powers nor short-lived periods of peace. This ensured that war-planning was not linked to securing such agreements. Furthermore, the burden of war was continual.

These aspects of the total conflict of 1914–45 do not mean that Napoleonic warfare was limited, certainly in terms of the conventions of the period. Given the resources available in a Europe that was still essentially a low-efficiency peasant society, Napoleonic warfare was an atrocious burden: more so, for example, than the Second World War was for the USA. Yet, in tactical and strategic terms, there were clearly limits compared to what came after. The Revolutionary and Napoleonic Wars saw experiments with submarine warfare, the first use of the air for conflict (reconnaissance balloons for artillery spotting), and the first use in the West of rockets,[28] but none of these made any real impact on the war, and investment in all of them combined was very small. It was not until the twentieth century that the dimensions of conflict expanded with air, submarine and rocket warfare. Furthermore, it was only then that the chemistry of war acquired the tool of gas.

These points cast into perspective the undoubted technical developments in battlefield capability and operations in the nineteenth century, and yet the latter were important. Land warfare was transformed by the continual incremental developments in firearms, from steps such as the introduction of the minié ball rifle and, subsequently, of breech-loading cartridge rifles. The net effect for both hand-held firearms and artillery was very substantial changes. In addition, there was greater and more predictable production of munitions from a more streamlined and systematized manufacturing process in the century after Napoleon's final defeat at Waterloo in 1815. The net result was a degree of change that was greater than that over the previous century.[29]

This was even more the case if logistics, command and control, and naval warfare were considered. Steam power,[30] the railway and the telegraph made a major difference for the Western powers, both to nearby operations, for example the mid-century wars in Europe[31] and the American Civil War, and to wars waged at a greater distance.[32] The combination of the three made it possible to apply and direct greater resources, and in a more sustained fashion, than hitherto. This did not necessarily determine the course of conflict with those who lacked such technology, but it did make it far easier to organize war. A comparison, for example, of the British capture of Manila in 1762[33] and the American conquest of the Philippines from 1898 reveals very different military systems.[34] So do the very limited French presence in Madagascar in the seventeenth and eighteenth centuries, [35] and their conquest of the island in 1894–5.[36] Technological advances made European operations in Africa easier in the 1880s and 1890s than they had been earlier.[37]

In Europe, the potential to apply more resources was linked to their greater availability, thanks to demographic growth, industrialization and militarization, and the utilization of these through effective systems of conscription, taxation and borrowing. The armed nation was more potent than ever before, thanks to the combination of citizen soldiers, nationalist motivation and industrialization. It was less clear how best to employ these forces, and this helps to account for interest in military science and theory,[38] as did the intellectualization of subjects in an age of increasingly self-conscious specialization and professionalization. The machinization of war gripped the imagination of many commentators as a consequence of, and force for, modernity and modernization.[39]

Yet it is also appropriate not to exaggerate the machinization of European warfare. When the First World War broke out, there was an average of only one machine-gun per thousand troops. The Serbs in part relied for their communications on 192 homing pigeons.[40] The far more numerous Russians were also poorly equipped in 1914, and a successful combination of government and industry to foster effective rearmament had not been achieved.[41] More generally, the creation of large conscript armies created serious problems of training, equipment and command.[42] These were not readily apparent in peacetime, but conflict found the military unable to operate as envisaged. Bulk did not equal mass, nor movement manoeuvre. Logistical support was frequently inadequate.

This was even true of forces generally regarded as highly efficient, such as the German army of the Second World War. The German invasion of the Soviet Union in 1941 suffered from serious problems with logistics and the speed of advance before the autumn rains and the onset of winter, unsurprisingly so, as its infantry depended on horse-drawn transport. Thus, seeing an increased size of the armed forces as necessarily correspondingly valuable as a wartime device is flawed, although size was clearly very useful in peacetime sabre-rattling. This problem of quantity was primarily one for land warfare, although it could also be an issue at sea.

The French Revolutionary and Napoleonic period left examples along the continuum from rapid and victorious offensives to slogging matches, a continuum that had also been true of *ancien régime* warfare, despite misleading subsequent attempts to present it as uniform, and also as limited and indecisive. Napoleon's 1805 campaign, with its victories over the Austrians at Ulm and the Austrians and Russians at Austerlitz, leading to Austria's departure from the war, and his 1806 triumph over Prussia, served as models for subsequent commentators, but this underrated different lessons that could be learned from his 1808 invasion of Spain, the 1809 war with Austria, and the 1812 invasion of Russia. The first indicated the limited value of battlefield victory if it did not lead to a surrender that prevented popular resistance. The second, at Wagram, highlighted the attritional possibilities of battle, and the campaign has been seen as prefiguring much in modern warfare.[43] The third indicated the same at Borodino, but also the difficulty of forcing both battle and political settlement.

The complex legacy of the Napoleonic wars was not, however, adequately teased out, not least because they were followed by over three decades without major conflict in Europe. The next major bout of warfare, the Crimean War and the Wars of Italian and

German Unification (1854–71), was waged with different weapons and with a role for new technology. Thus, it was important that the steamships that took French, British and Piedmontese troops to the Crimea in 1854–5 were not matched on the Russian side by railways capable of moving large numbers of troops and ample supplies. In contrast, in 1859, both Austria and France used railways in the mobilization and deployment of their forces. Alongside new technology, there was a widespread militarization of society, involving a propagation of military service as an honourable duty and of participation in war as glorious.

Again, the conflicts of the period led to potentially different conclusions. The one that received most attention focused on Prussian successes. These led to the conclusion that offensive operations that were carefully planned by an effective and professional general staff and that drew on the logistical possibilities provided by railways would lead to decisive victories. In 1866, the Prussian General Staff co-ordinated the concentration of three armies against Austrian forces. The prestige of the German military encouraged emulation, both within and outside Europe.

Yet Moltke, the architect of these campaigns, himself warned of the hazards of extrapolating from them. While arguing that it was preferable to fight on the territory of one's opponent, he was increasingly sceptical about the potential of the strategic offensive.[44] Prussian skill had not prevented many difficulties, not least at the hands of Austrian artillery and French *chassepot* rifles. Furthermore, deficiencies in leadership and strategy on the part of Austria[45] and France played into Prussian hands, enabling the Prussians to outmanoeuvre their opponents. This looked towards the danger of assuming that German successes in 1939 and 1940 demonstrated the ineluctable superiority of *Blitzkrieg*. In addition, in neither 1866 nor 1870 did Prussia have to conquer its opponents. To that extent, the wars more closely approximated to Napoleon's attacks on Austria than to his invasions of Spain and Russia.

Indeed, Prussia and the Prussian army were not really up to the task of conquering their opponents. After the army's successes in the decisive battles fought near the frontier in 1870, in which the major

French armies had been outmanoeuvred, the Prussians encountered difficulties as they advanced further, not least from a hostile population. More generally, the resources for the total conquest of Austria and later France, including, for example, possibly resisting a war of *revanche* mounted from the French colony of Algeria, were simply not present. This was not the option the government or Moltke sought anyway. Instead, they wanted a swift and popular conflict, with relatively low casualties.

Furthermore, European commentators did not draw the appropriate lessons from the American Civil War of 1861–5. Contemptuous of the Americans, they underrated the role of resources in the conflict, its length and, in part, attritional quality, the signs of what were to be seen as total war, and the development of trench tactics around Petersburg. The Civil War suggested that a major conflict would be protracted. Over one million died or were wounded in the war, in part because of the use of new firearms, particularly the percussion-lock rifle. Any protracted conflict would place a premium on the organization of resources.[46]

The difficulty of knowing what lesson to draw from the 1860s could be repeated by considering 1899–1905, for the Second Anglo–Boer War (1899–1902) and the Russo–Japanese War (1904–5) were seen as sending contrary messages, not least about the value of the offensive and how best to mount it, and the effects of artillery and small arms.[47] This is a warning today about the need to appreciate the complexity both of assessing present conflicts and the likely shape of future warfare. The Boer War received less attention because it seemed to have less relevance for European conflict, not least because it was a colonial war and fought with a low density of forces.

In contrast, the Russo–Japanese War involved very large regular forces and more recognizable front lines. As a consequence of the attritional character of the latter, the military anticipated heavy casualties in any future major war in Europe. They did not, however, appreciate how long such a conflict might be. The major lesson drawn from the Russo–Japanese War was that the Japanese, who had attacked, had won. In short, seizing the initiative, closing with the enemy, and troop morale seemed crucial, and it was possible to wage a modern war in the face of defensive firepower successfully and without the conflict taking too long. At the same time, the German Schlieffen Plan drawn up in the winter of 1905–6 by the outgoing Chief of the General Staff aimed for an even shorter conflict, by outmanoeuvring the French border defences and invading through Belgium.

1914–45

In the major wars of 1914–18 and 1939–45, the Germans sought to repeat Moltke's successes (and those earlier of Napoleon) by mounting and winning a war of manoeuvre. Moltke had adapted Napoleonic ideas of the continuous offensive to the practicalities of the industrial age, including railways. In place of frontal attack, he had sought to destroy the cohesion of the enemy army and to envelop opposing forces. This also was the German plan in 1914,[48] and – albeit with the addition of air power and tanks – 1939–40. The French would be denied the opportunity to stage a fighting retreat into their interior, and would thus lose the advantage of space.

Such a strategy was that of the military first strike, and this encouraged the outbreak of hostilities, as did the heightened sense, created by pre-war military preparations and a series of diplomatic crises, that war was likely.[49] German policy-makers looked back to Frederick the Great's surprise invasions of Silesia in 1740 and Saxony in 1756. The politics of first strike was further driven by the German fear of simultaneous conflicts with powers to west and east, specifically, in 1914, France and its ally Russia. It was dangerous, however, for two reasons. First, wars of aggression could provoke the very combination Germany feared. Indeed, they could give it added force. Thus, in 1914, the invasion of Belgium, a step taken in order to envelop the French from their left, provoked British entry into the war, ensuring that, if it became a long-term struggle, Germany would find it far harder to win.

Second, an emphasis on first strike and the maintenance of the initiative tended to ensure that there was insufficient preparation and planning for long-term struggle. This affected Germany in both World Wars, and Italy and Japan in the Second World War. In particular, there was insufficient thought given to two different problems: first, how, in the event of a long conflict, to obtain the maximum political, economic and military benefit from areas that had been overrun; and second, how to prepare for defensive war. German warmaking, with its emphasis on surprise, speed, and overwhelming and dynamic force at the chosen point of contact, was designed for an offensive strategy that was most effective against linear defences, not against defence in depth that retained the capacity to use reserves. The Japanese were not only over-confident and poorly prepared for a long struggle, but also suffered from a long-standing failure to reconcile army and naval objectives, priorities and methods.

The influential military historian Hans Delbrück argued in 1913 that a limited conflict was unlikely, and warned of the danger that European war would devastate society. By December 1914, he had concluded that a political solution to the war should be sought. After the First World War, Delbrück waged a bitter controversy with the military over German policy in the conflict. Ludendorff and his supporters condemned Delbrück (a civilian) as a journalist and an armchair strategist who lacked the intuitive understanding of war that came from technical training and practical military experience, while Delbrück charged the military leadership with disastrous ambition.[50]

German warmaking was not suited to defensive war, which was antithetical to the self-confidence and sense of mission of the Wilhelmine and Nazi regimes. Although both proved formidable foes on the defensive, they jeopardized the strength of their defensive situation, specifically their resource base, by fresh offensives, such as those on the Western Front in 1916 and 1918, and the Kursk offensive against the Soviet Union in 1943. Once the war of manoeuvre of the early stages of the conflict had been lost, it was difficult and costly to regain it, as the Germans discovered on the Western Front in 1918 and late 1944, and the Japanese, in a different way, in the Solomons, at Midway, and in Burma in 1944. Difficult, but not impossible, as the Germans showed across the greater distances of the Eastern Front both in the First World War and in 1942, and in North Africa in 1942, and the Japanese in southern China in 1944.[51] Furthermore, their renewed offensives were designed to reap political and military benefits that were worth seeking.

Yet German and Japanese failure was not simply a consequence of the replacement of the war of manoeuvre by one of attrition. It is also necessary to note the limitation of both their military machines for offensive operations, the flaws in strategy and command, and the fighting quality of their opponents, that ensured that they did not receive due benefit from taking the initiative. At Gembloux on 14–15 May 1940, the weaknesses of the *Blitzkrieg* tactical system were demonstrated by a French artillery-infantry defence, especially by the French artillery.[52] More generally, the Germans lacked sufficient mechanization for their infantry and logistics, adequate artillery support,[53] and sufficient effective ground-support air power. More seriously, they had not planned adequately for a sequence of campaigns and a long war. The space of the Soviet Union had not been conceptually overcome.

German and Japanese offensive strategies worked best if the opponent was poorly prepared to contest a rapid advance from a number of directions (Poland 1939, Yugoslavia 1941, Malaya 1941–2), or lacked a strategic reserve (Poland 1939, Yugoslavia 1941), or if the reserve was inadequate and/or poorly directed (France 1940, Malaya 1941–2). The Poles and the French placed the bulk of their forces too far forward.

If, in contrast, a reserve existed and was mobile and well handled, as was eventually the case in Russia in 1941, then the initial success of the offensive could be very difficult to sustain. Space was an aspect of the reserve, and, if properly handled, a force-multiplier for the defensive. This was shown in the Second World War in China, Russia, South Asia, the Pacific and North Africa, and, once it had been overrun, in Yugoslavia. In addition, there were tactical problems with mechanized assaults, not least the difficulty of maintaining infantry support. This was important for countering anti-tank guns, as the Israelis were also to discover in 1973.

Both major wars in part developed into attritional struggles in which the fighting quality of German and, in the Second World War, Japanese units could not compensate for the number of their opponents and their successful mobilization of greater resources. It is also necessary to note the improvement in the fighting quality and command performance of the Allies: for example, in the Second World War, of the Soviet forces,[54] of the British army in Burma, and, in terms of doctrine and experience, of anti-submarine operations in the Atlantic. The Japanese in Burma and the Pacific were far less able to gain from experience, in part because they continued to rely on *esprit* rather than on careful adaptation. Repeatedly, the Japanese, believing that success was largely a matter of will, employed the same tactics. Similarly, Japanese naval tactics have been seen as overly dogmatic.[55] Against mobile Soviet forces, the Japanese proved deficient both in 1939 and in 1945.

The technological developments in war, most prominently, although not only, tanks, air power and submarines (and their respective counter-weapons), contributed to the attrition, instead of preventing it. As initial advantages were checked, so further innovations were introduced, for example *Schnorkel* devices allowing submarines to recharge their batteries while submerged, and acoustic homing torpedoes by the Germans in the Second World War.

Furthermore, the politics of the Second World War led to the goal of unconditional surrender, which accentuated the attritional character of the conflict, as well as its atypicality. This goal was a response to the character of the Nazi regime, and serves as a reminder of the flawed political analysis of the German military leadership. Their willingness to accept Hitler not only morally corrupted them, as the military came to play a role in Hitler's genocidal policies, but also led them into a conflict in which, from 1941, limited war and political compromise ceased to be options. As a consequence, the operational ability of much of the German military was linked to a task that risked, and in the end caused, not only their defeat but also their dissolution.

However, to underline the role of counterfactual arguments, the alliances in each conflict were neither inevitable nor stable. This provided opportunities for opponents, and also obliged powers to take note of the views of allies when devising coalition strategies. Their ability or failure to do so successfully was a factor in determining victory.[56] This echoed earlier conflicts, for example the French treatment of Spain in the 1800s discussed in Chapter Seven, and also looked forward to the folly of discussing the likely purposes, nature and course of future conflicts without assessing the implications of alliance politics.

The terrible casualties of both World Wars have tended to make the conflicts appear more attritional than was in fact the case, although there were specific engagements which were like this. If the Germans and Japanese lost the ability to mount victorious campaigns of manoeuvre, that does not mean that their opponents simply wore them down. Instead, like the Confederates in the American Civil War, they were outfought, and defeated both in offensive and in defensive warfare. After the Germans, helped by Russian instability, had knocked Russia out of the war in 1917, they suffered the same fate the following year. Effective Allied, especially British, artillery–infantry co-ordination on the Western Front and the development of indirect (three-dimensional) firepower played a major role in a significant pushing back of the Germans, although it subsequently attracted less attention than the more novel use of tanks. A similar point can be made about British operations in Palestine in 1917–18, although cavalry, rather than tanks, received disproportionate attention in this case.[57] More generally, aircraft expanded the three-dimensional battlefield. On the Western Front, the adoption of unity of command under Foch in April 1918 also greatly helped the Allies.

However, the popular image of the First World War – of total futility and mindless slaughter – is such that it can claim to be the most misunderstood major conflict in history.[58] This image, in a way, is important to *Western* twentieth-century military history, as it reflects a disenchantment with war understood both as a pursuit of state interest and as fighting. Popular military history continues to strike a familiar note. For example, in his *The First World War* (1998), the prominent popular military historian John Keegan wrote of the British landing at V Beach in the Gallipoli campaign of 1915: 'The columns on the gangplanks, packed like cattle ranked for slaughter in an abattoir . . .'.[59] Of course, they were not intended for slaughter, nor to be eaten. The image is totally misplaced, but the very fact that such an inappropriate phrase can be employed is indicative of a wider failure to understand the conflict. To return to Keegan, on the final page of the text, he declares: '. . . the First World War is a mystery. Its origins are mysterious. So is its course' – remarks that are surprising given the wealth of scholarship on both. Again:

Why, when the hope of bringing the conflict to a quick and decisive conclusion was everywhere dashed . . . did the combatants decide nevertheless to persist . . . and eventually to commit the totality of their young manhood to mutual and existentially pointless slaughter . . . the principle of the sanctity of international treaty, which brought Britain into the war, scarcely merited the price eventually paid for its protection.[60]

As before, this tells us more about the values of a later age than those of the 1910s. Not that Keegan would go in that direction, but the anti-militarism his book possibly unintentionally echoes had, as a consequence, not only values of restraint that are often applauded, but also the defeatist collaboration that followed the German defeat of France in 1940. A sense of the obsolescence of warfare can be seen throughout Keegan's recent work, especially his *A History of Warfare* (1993) and *War and Our World* (1998), but this response to the potential of nuclear weaponry is a limited guide to earlier conflict and, arguably, also to the modern situation and the likely future.[61]

To return to the First World War, much of the campaigning was successful. The Austro-German-Bulgarian conquests of Serbia (1915) and Romania (1916) were decisive victories. The Germans defeated Russia, and in 1917 Austro-German forces routed the Italians at Caporetto, although the Italians subsequently held their opponents' advance at the Piave. The British conquest of Palestine from the Turks in 1917–18 displayed impressive strategic and operational flexibility.

The portrayal of the war in terms of the Western Front is thus limiting. The conflict in the west was very important, but it was not all the war, nor the sum total of strategic assessment and fighting methods on land. The casualty figures reflected not so much the futility of war, but the determination of the world's leading industrial powers to continue the war and, more specifically, the strength of counter-strategy and tactics, or, phrased differently, the value of defence in moderate depth, given the contemporary constraints of offensive warfare and the numbers of troops available for the defence.

Yet that did not prevent effective offensive tactics, particularly by both the Germans and then the British in 1918. Even before that, there had been successful attacks, as with the Allied capture of Messines Ridge in 1917, an artillery victory.[62] The 1918 German offensive emphasized surprise, tempo and tactical flexibility, with, for example, only a short preparatory artillery bombardment and a non-linear infantry advance. The Germans aimed for breakthrough not as an end result of heavy fighting, as in offensives on the Western Front in 1915–17, but as the immediate goal. This was successful on the battlefield, but the British were able to deploy reserves such that the German tactical success was not translated into strategic victory.[63] Nevertheless, the Germans had made major advances in the organization and use of their artillery and in their infantry tactics.[64]

The Allied offensive in 1918, especially that by British, Canadian and Australian forces, was not only effective, but also important to the development of modern warfare. In place of generalized firepower, there was systematic co-ordination, reflecting precise control of both infantry and massive artillery support and improved communications. The British army had 440 heavy artillery batteries in November 1918, compared to 6 in 1914.[65] The 1918 Allied campaign can be described as attritional, but also shows, as does the Soviet defence at Kursk in 1943,[66] that such campaigning could deliver a decisive result.

The successes of infantry–artillery co-operation in 1918 were overlooked because of what has been seen as the conceptual revolution of the inter-war period, and was seen in that light by many of its protagonists. Thus, in Britain, Germany, Russia, the USA and elsewhere, there was intellectual enquiry about the nature of war-winning, attempts to develop operational doctrine, concern about ensuring manoeuvrability and what would later be termed 'deep battle', and much interest in the potential of tanks, mechanized transport, air-to-land and air-to-sea warfare, and also in the enhanced communication capability offered by radio.[67] In practice, ideas advanced further than technological capability and resource availability, as was to be shown in the early land campaigns in the Second World War and in strategic bombing.[68]

Even Western states not associated with the development of new operational doctrine in the inter-war period were still affected by the introduction of new weaponry and still concerned to learn lessons from recent conflict – the last a characteristic of the Western military over the last two centuries. For example, in 1939 the Polish cavalry, about 10 per cent of the men under arms, was armed with anti-tank and heavy machine-guns, and trained to fight dismounted, the horses being employed in order to change position after an action – in short, for mobility, not shock attack.[69]

Yet the example of Poland also underlines the difficulty of deciding what lessons to draw from recent conflict, as well as indicating the role of military politics in the development of doctrine and organization. Polish victory over Russia in 1920, not least at Komarów, the last cavalry battle in Europe, led to a mistaken confidence in the continued value of the methods used then. The Poles did not match the mechanization of the German and Soviet armies in the 1930s. General Sikorski pressed the value of mechanized warfare and the tank, but he was out of favour with Marshal Pilsudski, who had staged a coup in 1926, and was dictator thereafter until his death in 1935 (the military dictatorship continued until Poland was conquered in 1939). Although, in the late 1930s, the Polish military came to understand the value of armour, they were too far behind their rivals.[70]

The fate of Poland in 1939 is therefore a demonstration of the failure to maintain technological capability, in this case specifically mobility, although other factors were also of importance, including the dry weather that permitted the Germans to benefit from this mobility. Polish dispositions were inappropriate, not least the strung-out nature of the defensive perimeter and the lack of defence in depth, but the politics of the conflict were also crucial. The Soviets joined the Germans, while Poland's Western allies did not mount diversionary attacks. Poland also suffered from the earlier destruction of Czechoslovakia – a destruction its government had been happy to share in.

Learning lessons and doctrinal innovation sounds far more clear-cut and easier than is, in fact, the case;[71] so also with the evaluation of the process. One particular problem concerns the assessment of 'anti'-strategies and tactics. If, for example, the doctrine and technology of a period favours the offensive, then investing in the defensive can be seen as anachronistic, or as a way to try to lessen the impact of the offensive. It can, indeed, be regarded as prescient, even forward-looking. The last is going too far in the case of inter-war French investment in a system of frontier fortifications – the Maginot Line[72] – but it is worth noting the extent to which this Line, and the pre 1914 fortifications, affected German strategy. Investment in fighter aircraft was a defensive response to the development of bombers, and a successful one, eventually forcing the creation of long distance escort fighter wings.

The Spanish Civil War of 1936–9 was a major struggle that offered the possibility of testing out new weapons and tactics and of learning military lessons. Germany, Italy and the Soviet Union played a role in the conflict, which other powers followed with close attention. However, the ability to learn lessons was limited by inter-service reality and the nature of air force culture, especially in Britain. In particular, there was resistance to the possibilities of ground-support operations on the part of those committed to strategic bombing, while the need to provide fighter escorts was not appreciated by the British or the Americans until after heavy casualties in the Second World War.[73] Lessons were also not drawn from the Sino–Japanese war which began in 1937. For example, the battle of Shanghai in 1937 showed that Chinese losses made during the day could be regained at night when Japanese air power was less potent.

In the Second World War, warfare on land in Europe was more similar to the situation in the First World War than is sometimes appreciated.[74] It was also decisive. In 1939–40, the Germans broke the threat of encirclement and war on two fronts, and the more likely danger that Britain and France would be able to rest successfully on the defensive behind the Maginot Line while blockading Germany into a change of policy or leadership. Later, the Soviets, in less than two-and-a-half years' fighting, drove the Germans from the Volga to the Elbe, a distance greater than that achieved by any European force for over a century, and certainly one that does not suggest that a war of fronts precluded one of a frequent movement of these fronts. Soviet operational art towards the end of the war stressed firepower, but also employed mobile tank warfare: attrition and manoeuvre were combined.[75] Similarly, the Western Allies drove the Germans from Normandy to the Elbe in less than a year.[76] As in 1918, there was no stab in the back: the Germans were beaten. As with the German offensives in 1939–40, the Allies destroyed their opponents' forces: war was decisive in this double sense.

As in the First World War, the high rate of depletion of men and resources on both sides that leads to the description attrition did not produce indecisive warfare. Nor did it preclude a major role for generalship, strategic skill, operational innovation and tactical fighting quality. They, as well as resources, were important; indeed, they represented the application of resources. Although the fighting was less continuous and the resources devoted by both sides far less, the same was also true of the Allied expulsion of the Germans from North Africa and the subsequent successful invasions of Sicily and mainland Italy.

As with the First World War, military history still has a major role to play for this much-studied conflict. In part, this reflects major lacunae in the documentation or scholarship, as – until recently –

with the war on the Eastern Front. In this case, it is possible to direct attention to important Soviet operations that failed and were subsequently ignored, for example OPERATION MARS in November–December 1942, the Soviet Central and Kursk Front offensive of February–March 1943, and the Belorussian operation of November–December 1943. It is also possible to re-evaluate positively the Red Army, in order to suggest that it was a more effective fighting force than German commentators allowed, and that its victory was not simply a consequence of resources and a willingness to take losses. Soviet doctrine, with its emphasis on defence in depth and its stress on artillery in defence and attack, proved more effective once the initial shock and surprise of the German attack had been absorbed. Once their advances had been held, the Germans suffered from the absence of sufficient manpower, artillery and supplies. Stabilizing, let alone advancing, the front proved to be an enormous strain on resources.[77] Similarly, there has been inadequate coverage of Japanese operations in 1941–2 and what they have to contribute to the understanding of amphibious warfare.[78] In part, as with these examples, or that of the Western Front in the First World War, it is also necessary to test widely held assumptions.

Given the emphasis in Chapter Seven on the contingent in military history and international relations, it is worth considering whether the two World Wars and the third great, but different, conflict of the century, the Cold War, can be assessed in the same way. This involves asking whether the role of resources overcomes the contingent, if not in particulars, at least in the long term and overall; in short, were the Germans and, in the Cold War, the Soviets bound to lose? This is an issue that divides scholars.[79]

It would be absurd to ignore the role of resources, of troops, material and funds, as well as the more basic resources of population, raw materials and industrial plant. These resources were mobilized to a formidable extent, producing unprecedented quantities. In addition, these resources were utilized with a novel degree of scientific application. This was institutionalized in bodies such as the Office of Scientific Research and Development in the

USA. The relationship between such research and military development continued after the war. In 1945, for example, Theodore von Kármán, an American Air Force Science adviser, was responsible for heading a long-term scientific forecast on behalf of the air force. This process was replicated in war-related industries.

Furthermore, the mobilization of resources involved a marked degree of direction of the economy. This was particularly true of the Soviet Union. Already an autocracy, where economic planning and the brutalization of society were mutually supporting, the Soviet Union lost many of its leading agricultural and industrial areas to German advances in 1941 and 1942. Manpower, however, still existed, and the Soviet Union was able to deploy vast forces and to take heavy casualties, including at least 9.6 million dead troops,[80] and, in addition, civilian volunteer militia and partisans.

Problems of resource availability and utilization affected even the strongest state – in the Second World War, the USA – and the role of political cultures and institutional practices in determining the availability and use of resources should not be underrated. For example, the major German company Daimler-Benz was unwilling to commit itself too heavily to armament production before and during the Second World War, as war would lead to victory, or, later, defeat, and in either case it was necessary for the company to plan for peace. More generally, the weakness of government direction hit German armament production.[81] In the Vietnam War, American strategy was constrained by resource shortages,[82] although, for political reasons, the Americans did not mobilize their resources fully.

It is instructive to note the enormous economic and social damage imposed on countries such as the Soviet Union[83] and Britain in the Second World War. The availability of vast resources did not overcome the contingent in the short term, and the long term is really only a series of short terms. Looked at differently, the contingent fact of war helped mould the long-term resource base, socio-economic structures and political cultures of the combatants.

Equally, the course of the conflicts included multiple contingencies and counterfactuals: from the grand – what if Britain had not entered the First World War or Hitler had not declared war on the USA or there had been no Sino–Soviet split – to the less powerful, but still important. To be more specific, it can be suggested, for example, that, alongside the view that Japanese success helped ensure the redundancy of European imperial strategies in East Asia and, crucially, destroyed European prestige, it is also possible to focus on the political and resource costs of the European wars of 1914–45. In short, did Britain's decision to abandon its colonial presence in India stem, in part, from conflict *within* the Western system? In the case of the USA, it was far from clear how far it would be committed to the two World Wars, while there were also major differences over the strategy that should be followed in the late 1940s and 1950s, not least the degree to which the USA would follow a policy of global containment of Communism and would rely on nuclear weaponry, and thus air power.[84]

Each counterfactual can be variously debated, but their net effect is to take us towards what should be a more tentative understanding of present and future. This can be seen not only with the two World Wars, but also with the Cold War, because it is a reminder of another important aspect of military history: namely, that the conflicts that did not occur (and, therefore, also those that do not and will not) are as important as those that did. This is true for empirical reasons: if x had used up many of its resources in conflict y, then it would have been less well placed for its next war, or conversely, might have had more military industrial capacity, political commitment to intervention, battle-ready forces and experienced commanders. Thus, the Americans were better able to fight the Korean War (1950–3) because of their role in the Second World War, although, meanwhile, their fighting effectiveness had declined, as was shown by the experience of some American units in the first year of the war. Even when their fighting quality improved, the Americans found that the strength and determination of the North Korean and Chinese responses were able to fight them to a stalemate.

Conflicts that did not occur were, and are, also significant for analytical purposes, given that judgements of the capability of military systems have to take note of all possible conflicts. Furthermore, conflicts that did not take place might well have led to developments in the science of war. For example, the absence, other than in very limited cases (although still appalling in their consequences), of atomic, bacteriological, chemical and aerial-borne gas attacks,[85] or of the large-scale use of rockets with sophisticated guidance systems (as opposed to V2s) is important to the norms of warfare at the start of the twenty-first century. The absence of major insurrectionary struggles (as opposed to terrorism) within the West, certainly in North America and Europe, is also important to the political dimensions of warfare and the military, as it ensures that organizational preparations for such conflict have been less intensive than they would otherwise have been.

AFTER 1945

The post-1945 period saw a major shift in the relations between the West and the non-West. This shift exposed the limitations of military technology, especially in the initial stages in which the authority of colonial powers was resisted by non-state/governmental movements, as with the resistance to French rule in Vietnam (1946–54) and Algeria (1954–62). This warfare tested Western militaries and the practices of colonial rule, and suggested the precarious nature of the latter in a modern context. Colonial conflict after 1945 also challenged Western militaries that had recently been forced in the Second World War to focus on war between conventional forces. This proved a major problem for the French, who lacked any recent experience in counter-insurgency operations (and who had also taken a very minor role in the conflict with Japan). The resources devoted were extensive. French forces in Algeria rose from about 50,000 in early 1955 to 620,000 in 1956, after first reservists and then conscripts were sent. The dispatch of both of these groups was unpopular, and greatly increased opposition to the conflict within France.

However, it is important not to adopt an overly schematic, even teleological, account and to minimize the earlier difficulties affecting colonial rule. For example, the problems the British confronted in India in the inter-war period in part looked back to the nineteenth century. In colonies such as India and Jamaica, administrators and officers had for a long time been concerned as to how best to control populous territories with very small forces. Military strength was seen as the crucial support of moral authority. The extent to which this authority was accepted by the colonized in the nineteenth century should not be exaggerated, but there is little doubt that it was far more under challenge by the inter-war years. There were serious problems in a number of colonies and positions, including Egypt, India and Palestine. In China, the British abandoned their concessions in Hankou and Jiujiang in 1927 after massive public protests overawed the local British military presence. A Chinese trade unionist was killed in each city by British troops, but whereas in 1925 the position in Hankou had been underpinned by local warlords, in 1927 there was no such backing.[86]

The situation was different when colonial rule was replaced by new states, a process that was largely (although not totally) complete in Africa, South Asia and the Middle East by 1976. The last British troops left Malaya in 1957 and Singapore in 1975, and the Portuguese coup of 1974 was followed by the withdrawal of Portuguese troops from colonies that were granted independence. Conflict between Western powers and newly independent states, and also between these newly independent states and insurrectionary movements, was different in character to that between Western colonial powers and insurrectionary movements. In the first case, conflict between Western powers and new states both involved regular forces on each side and provided the Western power with the military target of a governmental structure, and also with a political agency with which peace negotiations (and thus an agreed verdict to the conflict) were possible.

The prime difference between these conflicts and those of the age of Western imperialism was that now territorial conquest was not the objective. Thus, in 1936, Italy had conquered Ethiopia at the close of a war that it had started, but the Korean War was begun in 1950 by North Korea, and not fought by an American-led United

Nations coalition in order to create an American or UN colony. Again, in the Vietnam War, although the Americans sought to destroy an insurrectionary movement in South Vietnam, they did not seek to conquer the North and, instead, recognized its government.

Conflict between Western powers and newly independent states involved the Western powers in war with regular forces, most obviously in the Gulf War of 1990–1 with Iraq. This could be planned and prepared for in terms of the symmetrical warfare envisaged if the Cold War became hot. In practice, however, there were particular military and political circumstances and parameters in each confrontation or conflict. It proved necessary to learn this, in order to succeed in using force to obtain the desired political outcome. Most conflict between Western powers and independent states did not, however, entail full-scale hostilities. Instead, covert action, particularly to overthrow a government, was more common, as, for example, with American pressure on Nicaragua in the 1980s. The extent to which such action is seen as war varies.

Conflict between newly independent, or at least Third World (for China and Iran are not newly independent) states also rose as Western colonial rule receded, and the process was to be repeated in the 1990s after the collapse of Soviet power. Thus, for example, the ending of British and French rule in the Middle East was followed by a series of wars between Israel and its Arab neighbours (1948–9, 1956, 1967, 1973 and 1982). Conflicts between India and Pakistan followed the end of British rule in South Asia (1947–9, 1965 and 1971), while, in East Africa, Tanzania invaded Uganda in 1979. The end of French rule in Indo-China was followed by conflict between the successor states, at first between North and South Vietnam, but eventually also involving Laos and Cambodia. The end of British power in South Asia enabled China to challenge the McMahon Line which the British had imposed as the Indian–Tibetan border in 1913–14. This led to conflict between India and China in 1962.[87]

In the Caucasus, after the fall of the Soviet Union, Armenia fought Azerbaijan over Nagorno-Karabakh, while the Russian-backed attempts by Abkhazia and South Ossetia to secede from Georgia led to conflict. Elsewhere in the Former Soviet Union, the 'Trans Dniestr Republic' supported by former Soviet forces sought to break away from Moldova. Conflicts such as that between Vietnam and Cambodia in 1978–9, or clan-based civil war in Tajikistan in 1992–7 mirrored the rivalries and wars that had preceded and frequently facilitated European imperialism. As with the earlier conflicts, they also assisted the spread of Western influence.

These conflicts between Third World powers were, at times, an extension of the Cold War. This owed much to the search for allies on the part of the Western powers, the Eastern bloc, and rivals elsewhere in the world, and also to their anxieties. Weaponry, training and military advisers were provided by the Western powers and the Eastern bloc.[88] Thus, for example, after the Arab–Israeli Seven Day War in 1967, the combatants were rearmed by their patrons. This affected the military balance, especially with the Soviet installation of a powerful air-defence missile system for the Egyptians. Vast amount of munitions, allegedly $50 billion worth, were sold to Iraq during the 1980–8 conflict with Iran. In 1985, Britain signed a £20 billion arms contract with Saudi Arabia.

Conflicts in the Third World between regular forces could be analysed in familiar terms. For example, possibly in an overly self-serving fashion, Liddell Hart presented Israeli strategy and tactics in the Six Days' War of 1967 as an example of his theories and a continuation of the *Blitzkrieg* employed by the Germans, particularly in 1940.[89] Yet again, such a search for parallels could be misleading, as it underrated the potential role of other military and political factors. In this particular case, the role and capabilities of air power in 1967 in the Middle East were far greater than in France in 1940 (although they were insufficient to secure military and, still more, political objectives in Vietnam).

Conflict between newly independent states and insurrectionary movements became more common as the number of the former rose. It is, however, possibly the most obscure aspect of modern military history. Many of these struggles have not been fully studied, and their part in the role of the military across much of the world has been underrated. These struggles reflected the difficult legacy of colonial rule. This included frontiers bisecting peoples – a major problem in much of the world, especially much of Africa, and one that had indeed caused, or encouraged conflict in Europe, especially in 1830–1918. Thus, in 1999, the Lozi-speaking people of Caprivi sought to secede from Namibia. Furthermore, post-colonial states frequently lacked any practice of tolerance towards groups and regions that were outside the state hierarchy. The use of state power encouraged a violent response. This was a major problem in countries that suffered severe civil wars, such as Nigeria, Ethiopia, Angola, Rwanda and Zaire.[90]

In general, resulting conflicts opposed the regular forces, supported by the security services, and often by para-militaries drawing on groups that supported the government, to insurrectionary forces, some of which were of considerable strength. The latter could benefit from foreign assistance, as in the unsuccessful guerrilla war in Darfur against Omani rule in 1970–5, or might, as in Angola, have a long pedigree dating from a role in the anti-colonial struggle, and might also reflect the proto-statehood of the insurrectionary movement. This was true, for example, of the Kurds in Iraq, Iran and Turkey. Their struggle took on the character of full-scale campaigns, as in 1996, when Iraqi forces captured the Kurdish capital Arbil.

Foreign assistance could lead such struggles to flare up into state-to-state hostilities. Thus, in 1965 and 1999, Pakistani support for Kashmiri insurgents led to fighting with India, while South African support for opposition by UNITA to the Cuban-backed Angolan government in the 1980s led to large-scale conflict. Such intervention continues. For example, the Congolese civil war in 1998–9 saw Uganda and Rwanda supporting competing rebel factions, while Zimbabwe, Angola, Chad and Namibia backed President Kabila. The conflict between Eritrea and Ethiopia spilled over into the civil conflict within Somalia. Foreign assistance could also come from foreign imperial powers, generally for the successor state, not the insurgent movement. Examples included British backing for Oman in the 1970s, and French for a host of African states, including Chad, in the 1970s and 1980s.

The extent of violent opposition to newly independent states was overshadowed in the first three decades after the Second World War by the wars of decolonization and by the Korean and Vietnam conflicts. Thereafter, however, conflict between newly independent states and between these states and insurrectionary movements became more frequent, although it continued to receive insufficient attention, certainly relative to conflict or preparations for conflict involving the major powers. In turning to war today, it is important to try to correct this emphasis.

NOTES

1. See, for example, W.H. McNeill, *The Pursuit of Power: Technology, Armed Force, and Society Since AD 1000* (Oxford, 1983).
2. V.C. Purcell, *The Boxer Uprising* (Cambridge, 1963).
3. A modern edition, with an introduction by Douglas Porch, was published by the University of Nebraska Press in 1996.
4. B.A. Elleman, *A History of Modern Chinese Warfare: 1795–1989* (2000).
5. J. Needham, *Military Technology: The Gunpowder Epic* (Cambridge, 1987), pp. 392–8; J. Waley-Cohen, 'China and Western Technology in the Late Eighteenth Century', *American Historical Review*, 98 (1993), pp. 1531–2.
6. S. Shaw, 'The Origins of Ottoman Military Reform: The *Nizam-i Cedid* Army of Sultan Selim III', *Journal of Modern History*, 37 (1965), pp. 291–5.
7. See, for example, R.F. Hackett, *Yamagata Aritomo in the Rise of Modern Japan, 1838–1922* (Cambridge, MA, 1971); J. Dunn, 'Napoleonic Veterans and the Modernization of the Egyptian Army, 1817–1840', *Consortium on Revolutionary Europe: Proceedings 1992* (1993), pp. 468–75; B. Langensiepen and A. Gulcryuz, *The Ottoman Steam Navy, 1828–1923* (Aldershot, 1995).
8. R.J. Smith, *Mercenaries and Mandarins: The Ever-Victorious Army in Nineteenth-Century China* (New York, 1978).
9. S. Spector, *Li Hung-chang and the Huai Army: A Study in Nineteenth-Century Chinese Regionalism* (Seattle, WA, 1964). Difficulties in Persia emerge clearly in R. Matthee, 'Firearms. History', in E. Yarshater (ed.), *Encyclopaedia Iranica* IX (New York, 1998), pp. 623–5.
10. T.L. Kennedy, *The Arms of Kiangnan: Modernization in the Chinese Ordnance Industry* (Boulder, CO, 1978).
11. T. Kuo and L. Kwang-Ching, 'Self-Strengthening: The Pursuit of Western Technology', in J.K. Fairbank (ed.), *The Cambridge History of China X* (Cambridge, 1978), pp. 491–542. On the dockyard, see D. Pong, *Shen Pao-chen and China's Modernization in the Nineteenth Century* (Cambridge, MA, 1993).

12. E.S.K. Fung, *The Military Dimension of the Chinese Revolution: The New Army and Its Role in the Revolution of 1911* (Vancouver, 1980). I have benefited from the advice of Bill Roberts.

13. J. Ch'en, *Yuan Shih-k'ai, 1859–1916: Brutus Assumes the Purple* (2nd edn, 1972); E.P. Young, *The Presidency of Yuan Shih-k'ai: Liberalism and Dictatorship in Early Republican China* (Ann Arbor, MI, 1977).

14. A.J. Birtle, 'The U.S. Army's Pacification of Marinduque, Philippine Islands, April 1900–April 1901', *JMH*, 61 (1997).

15. P.S. Thompson, *The Natal Native Contingent in the Anglo–Zulu War, 1879* (Pietermaritzburg, 1997).

16. R.A. Johnson, *The Penjdeh Crisis and its Impact on the Great Game and the Defence of India, 1885–1897* (PhD, Exeter, 1999). For the problems facing the British, T.R. Moorman, *The Army in India and the Development of Frontier Warfare, 1849–1947* (1998).

17. T. Winichakul, *Siam Mapped: A History of the Geo-body of a Nation* (Honolulu, HI, 1994).

18. J.K. Noyes, *Colonial Space: Spatiality in the Discourse of German South-West Africa, 1884–1915* (Chur, 1992); T.J. Bassett, 'Cartography and Empire Building in Nineteenth-Century West Africa', *Geographical Review*, 84 (1994), pp. 316–35.

19. H.M. Ward, *The War of Independence and the Transformation of American Society* (1999).

20. E.J. Zürcher, *Arming the State. Military Conscription in the Middle East and Central Asia, 1775–1925* (1999). This is mostly on the Ottoman empire, although Egypt and Persia are also considered.

21. E.L. Rogan, *Frontiers of the State in the Late Ottoman Empire* (Cambridge, 1999), pp. 9–12 and *passim*; H.G. Marcus, *The Life and Times of Menelik II: Ethiopia, 1844–1914* (Oxford, 1975).

22. T. Yuan, 'Yakub Beg (1820–1877) and the Moslem Rebellion in Chinese Turkestan', *Central Asiatic Journal*, 6 (1961), pp. 134–67; L.B. Fields, *Tso Tsung'-t'ang and the Muslims: Statecraft in Northwest China, 1868–1880* (Kingston, Ont., 1978).

23. E. Mawdsley, *The Russian Civil War* (Boston, MA, 1987); M. Broxup, 'The Last *Ghazawat*: The 1920–1921 Uprising', in M. Broxup (ed.), *The North Caucasus Barrier* (1992), pp. 112–45.

24. For another example, see I.C.Y. Hsu, 'The Great Policy Debate in China, 1874: Maritime Defense vs. Frontier Defence', *Harvard Journal of Asiatic Studies*, 25 (1965), pp. 212–28.

25. M.C. Finley, *The Most Monstrous of Wars: The Napoleonic Guerrilla War in Southern Italy, 1806–11* (Columbia, SC, 1994); D.W. Alexander, *Rod of Iron: French Counterinsurgency Policy in Aragon during the Peninsular War* (Wilmington, DE, 1985); J.L. Tone, *The Fatal Knot: The Guerrilla War in Navarre* (Chapel Hill, NC, 1994).

26. S. Chiang, *The Nian Rebellion* (Seattle, WA, 1954); E.J. Perry, *Rebels and Revolutionaries in North China, 1845–1945* (Stanford, CA, 1980).

27. T. Lai et al., *A Tragic Beginning: The Taiwan Uprising of February 28, 1947* (Stanford, CA, 1991).

28. A. Roland, *Underwater Warfare in the Age of Sail* (Bloomington, IN, 1978); F. Winter, *The Golden Age of Rockets: Congreve and Hale Rockets of the Nineteenth Century* (Washington, DC, 1991).

29. J.J. Farley, *Making Arms in the Machine Age: Philadelphia's Frankford Arsenal, 1816–1870* (University Park, Maryland, 1994).

30. A. Lambert, *Steam, Steel and Shellfire: The Nineteenth-Century Naval Technical Revolution* (1992). For the limitations of pre-steam naval technology and thus capability, see J.H. Pryor, *Geography, Technology, and War: Studies in the Maritime History of the Mediterranean* (2nd edn, Cambridge, 1992).

31. D. Showalter, *Railroads and Rifles: Soldiers, Technology and the Unification of Germany* (Hamden, CT, 1975).

32. D.R. Hendrick, *The Tools of Empire: Technology and European Imperialism in the Nineteenth Century* (Oxford, 1981).

33. N. Tracy, *Manila Ransomed: The British Assault on Manila in the Seven Years War* (Exeter, 1995).

34. B.A. Linn, *The U.S. Army and Counterinsurgency in the Philippine War, 1899–1902* (Chapel Hill, NC, 1989).

35. A. Ray, 'France in Madagascar, 1642–1674', *Calcutta Historical Journal*, 6 (1982), pp. 33–63.

36. Y.G. Paillard, 'The French Expedition to Madagascar in 1895', in J.A. de Moor and H.L. Wesseling (eds), *Imperialism and War: Essays on Colonial Wars in Asia and Africa* (Leiden, 1989), pp. 168–88.

37. B. Vandervort, *Wars of Imperial Conquest in Africa, 1830–1914* (1998).

38. A. Gat, *The Origins of Military Thought from the Enlightenment to Clausewitz* (Oxford, 1989), and *The Development of Military Thought: The Nineteenth Century* (Oxford, 1993); C. Bassford, *Clausewitz in English: The Reception of Clausewitz in Britain and America, 1815–1945* (Oxford, 1994).

39. D. Pick, *War Machine: The Rationalisation of Slaughter in the Modern Age* (New Haven, CT, 1993).

40. J.M.B. Lyon, '"A Peasant Mob": The Serbian Army on the Eve of the Great War', *JMH*, 61 (1997), p. 493.

41. P. Gatrell, *Government, Industry and Rearmament in Russia, 1900–1914: The Last Argument of Tsarism* (Cambridge, 1994). For a more positive view, see B.W. Menning, *Bayonets Before Bullets: The Imperial Russian Army, 1861–1914* (Bloomington, IN, 1993).

42. See, for example, R.R. Reese, *Stalin's Reluctant Soldiers: A Social History of the Red Army, 1925–1941* (Lawrence, KS, 1996).

43. R.M. Epstein, *Napoleon's Last Victory and the Emergence of Modern War* (Lawrence, KS, 1994).

44. G. Rothenberg, foreword to D.J. Hughes (ed.), *Moltke on the Art of War* (Novato, CA, 1993), p. ix. See, more generally, A. Bucholz, *Moltke, Schlieffen, and Prussian War Planning* (1991) and *Moltke and the German Wars, 1864–71* (2001).

45. G. Wawro, *The Austro–Prussian War: Austria's War with Prussia and Italy in 1866* (Cambridge, 1996).

46. E. Hagerman, *The American Civil War and the Origins of Modern Warfare: Ideas, Organization, and Field Command* (Bloomington, IN, 1988).

47. B. Nasson, *The South African War 1899–1902* (1999); G. Wawro, *Warfare and Society in Europe 1792–1914* (2000), pp. 145–6.

48. G. Rothenberg, 'Moltke, Schlieffen, and the Doctrine of Strategic Envelopment', in P. Paret (ed.), *Makers of Modern Strategy* (Princeton, NJ, 1986), pp. 296–325.

49. D.G. Herrmann, *The Arming of Europe and the Making of the First World War* (Princeton, NJ, 1996); D. Stevenson, *Armaments and the Coming of War: Europe, 1904–1914* (Oxford, 1996).

50. A. Bucholz, *Hans Delbrück and the German Military Establishment* (Lincoln, NB, 1997).

51. H. Ch'i, *Nationalist China at War: Military Defeats and Political Collapse, 1937–1945* (Ann Arbor, MI, 1982).

52. J.A. Gunsberg, 'The Battle of Gembloux, 14–15 May 1940: The *"Blitzkrieg"* Checked', *JMH*, 64 (2000), pp. 97–140. D.M. Glantz (ed.), *The Initial Period of War on the Eastern Front, 22 June–August 1941* (1993).

53. J. Bailey, *Field Artillery and Firepower* (Oxford, 1989), pp. 209–27.

54. For one example, see C. Van Dyke, *The Soviet Invasion of Finland, 1939–1940* (1997).

55. D.C. Evans and M.R. Peattie, *Kaigun: Strategy, Tactics and Technology in the Imperial Japanese Navy, 1887–1941* (Annapolis, MD, 1997).

56. For a possibly overly pro-Italian view of the pernicious consequences of the German failure to co-operate with Italy in the Second World War, see J.J. Sadkovich, 'German Military Incompetence Through Italian Eyes', *War in History*, 1 (1994), pp. 61–2.

57. J. Bailey, *The First World War and the Birth of the Modern Style of Warfare* (Camberley, 1996), for example p. 3. See also R. Paschall, *The Defeat of Imperial Germany 1917–18* (Chapel Hill, North Carolina, 1989); T. Travers, *How the War Was Won. Command and Technology in the British Army on the Western Front 1917–1918* (1992); P. Griffith, *Battle Tactics of the Western Front: The British Army's Art of Attack 1916–18* (New Haven, CT, 1994); C. Johnson, *Breakthrough: Tactics, Technology and the Search for Victory on the Western Front in World War I* (Novato, CA, 1994); M. Hughes, 'General Allenby and the Palestine Campaign, 1917–18', *Journal of Strategic Studies*, 19/4 (1996), pp.

59–88. For an account of well-executed attacks by British divisions in April 1917, see K.W. Michinson, *Villiers–Plouich: Hindenburg Line* (1999).

58. Thoughtful responses to the dominant interpretation include R. Prior and T. Wilson, 'Paul Fussell at War', and D. Englander, 'Soldiering and Identity: Reflections on the Great War', *War in History*, 1 (1994), pp. 63–80, 300–18.

59. J. Keegan, *The First World War* (1998, 1999 edn), p. 264.

60. Ibid., p. 456.

61. For a clear-cut statement of other troubling aspects of Keegan's work, see C. Bassford, 'John Keegan and the Grand Tradition of Trashing Clausewitz: A Polemic', *War in History*, 1 (1994), pp. 319–36.

62. I. Passingham, *Pillars of Fire: The Battle of Messines Ridge, June 1917* (Stroud, 1999); J.P. Harris, *Amiens to the Armistice: The BEF in the Hundred Days' Campaign, 8 August–11 November 1918* (1998).

63. M. Middlebrook, *The Kaiser's Battle. 21st March 1918: The First Day of the German Spring Offensive* (1978).

64. D.T. Zabecki, *Steel Wind: Colonel Georg Bruchmüller and the Birth of Modern Artillery* (Westport, CT, 1994); B. Gudmundsen, *Stormtroop Tactics: Innovation in the German Army 1914–1918* (New York, 1989).

65. S.B. Schreiber, *Shock Army of the British Empire: The Canadian Corps in the Last 100 Days of the Great War* (Westport, CT, 1997).

66. W.S. Dunn, *Kursk: Hitler's Gamble, 1943* (Westport, CT, 1997).

67. R. Simpkin, *Deep Battle: The Brainchild of Marshal Tukhachevskii* (1987); H.R. Winton, *To Change an Army: General Sir John Burnett-Stuart and British Armoured Doctrine, 1927–1938* (Lawrence, KS, 1988); B.H. Reid, *Studies in British Military Thought: Debates with Fuller and Liddell Hart* (1998); R.M. Citino, *The Path to Blitzkrieg: Doctrine and Training in the German Army, 1920–1939* (Boulder, Col., 1999).

68. S.L. McFarland, *America's Pursuit of Precision Bombing, 1910–1945* (Washington, DC, 1995).

69. A. Suchcitz, 'Poland's Defence Preparations in 1939', in P.D. Stachura (ed.), *Poland Between the Wars, 1918–1939* (Basingstoke, 1998), pp. 109–10.

70. P.D. Stachura, 'The Battle of Warsaw, August 1920, and the Development of the Second Polish Republic', in P.D. Stachura (ed.), *Poland Between the Wars*, pp. 54–5.

71. B. Bond, *British Military Policy Between the Two World Wars* (Oxford, 1980); W. Murray and A.R. Millett (eds), *Military Innovation in the Interwar Period* (Cambridge, 1996).

72. A. Kemp, *The Maginot Line; Myth and Reality* (1981).

73. J.S. Corum, 'The Spanish Civil War: Lessons Learned and Not Learned by the Great Powers', *JMH*, 62 (1998), pp. 313–34.

74. G.D. Sheffield, '*Blitzkrieg* and Attrition: Land Operations in Europe 1914–45', in C. McInnes and G.D. Sheffield (eds), *Warfare in the Twentieth Century* (1988), pp. 51–79.

75. C. McInnes, *Men, Machines and the Emergence of Modern Warfare 1914–1945* (Camberley, 1992), p. 39.

76. R.F. Weigley, *Eisenhower's Lieutenants: The Campaign of France and Germany 1944–1945* (Bloomington, IN, 1981).

77. D.M. Glantz, 'The Red Army at War, 1941–1945: Sources and Interpretations', *JMH*, 62 (1998), pp. 595–617. See also Glanz and J. House, *When Titans Clashed – How the Red Army Stopped Hitler* (Lawrence, KS, 1995), and A. Hill, 'Recent Literature on the Great Patriotic War of the Soviet Union 1941–1945', *Contemporary European History*, 9 (2000), pp. 187–97.

78. M.J. Grove, 'The Development of Japanese Amphibious Warfare, 1874 to 1942', in G. Till, T. Farrell and M.J. Grove, *Amphibious Operations* (Camberley, 1997), p. 23. For the limitations of Japanese air power, A.D. Harvey, 'Army Air Force and Navy Air Force: Japanese Aviation and the Opening Phase of the War in the Far East', *War in History*, 6 (1999), pp. 174–204.

79. For example, see G. Raudzens, 'War-winning Weapons: The Measurement of Technological Determinism in Military History', *JMH*, 54 (1990), pp. 403–33; M.E. Porter, *The Competitive Advantage of Nations* (1990); R. Overy, *Why the West Won* (1995).

80. G.F. Krivosheev, *Soviet Casualties and Combat Losses in the Twentieth Century* (1997).

81. N. Gregor, *Daimler-Benz in the Third Reich* (New Haven, CT, 1998).

82. R.H. Collins, 'The Economic Crisis of 1968 and the Waning of the "American Century"', *American Historical Review*, 101 (1996).

83. M. Harrison, *Accounting for War: Soviet Production, Employment, and the Defence Burden, 1940–45* (Cambridge, 1996).

84. M.J. Hogan, *A Cross of Iron: Harry S. Truman and the Origins of the National Security State, 1945–1954* (Cambridge, 1998).

85. S.H. Harris, *Factories of Death: Japanese Biological Warfare 1932–1945 and the American Cover-up* (1994).

86. D.G. Boyce, 'From Assaye to the *Assaye*: Reflections on British Government, Force, and Moral Authority in India', *JMH*, 63 (1999), pp. 643–68; S.K. Fung, *The Diplomacy of Imperial Retreat: Britain's South China Policy, 1924–1931* (Oxford, 1991).

87. N. Maxwell, *India's China War* (1970).

88. S. Green, *Living by the Sword: America and Israel in the Middle East 1968–1987* (1988); K. Krause, *Arms and the State: Patterns of Military Production and Trade* (Cambridge, 1995); A. Ilan, *The Origin of the Arab–Israeli Arms Race* (Basingstoke, 1996).

89. See, more generally, J. Mearsheimer, *Liddell Hart and the Weight of History* (Ithaca, NY, 1988).

90. J. de St. J. Jorré, *The Nigerian Civil War* (1972).

NINE

Conflict Today

We must not forget that war has its political side, and that the
sentiments of the people have to be considered.

General Hutchinson, 1905.[1]

As with earlier chapters, the emphasis in this one is on the
purposes of conflict and the organization of the military, rather
than on the details of weaponry. Most conflict today, understood as
the current situation and that over the last two decades, occurs in
the Third World; but it is also necessary to consider the changing
position of the military in the major powers. In organizational
terms, as discussed in Chapter Three, this is a matter of
professionalization and the depoliticization of the military. Both can
be seen in the world's leading military power, the USA, but also not
only there. The British example is instructive because British forces
have been involved in internal policing (in Northern Ireland)
throughout the period, as well as in foreign conflicts, both on their
own (the Falklands, 1982), and as a part of coalition forces (Gulf
War, 1991; Kosovo, 1999). More generally, the British example
indicates the possibly tentative nature of the professionalization and
depoliticization that are so widely stressed. Consideration of the
British case leads into discussion of the RAM (Revolution in
Attitudes to the Military), and then into the situation around the
world.

Force has a political character and context. It did not take
Clausewitz to discover this, but the relationship is one that has to be
reconsidered each generation. The nature of the public discussion of
this relationship is itself part of the equation, for the ability to

237

discuss military purposes is not a constant. For much of British history, military purpose has been a controversial issue. This was true of the size, composition, control and intentions of the armed forces, whether in the early thirteenth century or in 1641, in the 1680s or the late 1810s. These debates have been closely related to, indeed frequently coterminous with, controversies over the constitution and over foreign policy.[2]

The situation during recent decades in Britain has been different. There has been a strong degree of bipartisan consensus over defence, foreign policy and the constitution. This was expressed in a number of mutually reinforcing ways, so that such bipartisanship became normative. This was true of the formation of the wartime coalition government in 1940, and the suppression of party politics until after peace in Europe in 1945; of the formation of NATO in 1949, and its subsequent maintenance by both Labour and Conservative governments, and of the support of both governments for the nuclear deterrent and its upgrading, for the deployment of troops in Northern Ireland, and for membership of the European Union.

This political context greatly lessened the possibility of tensions in government–military relations,[3] but it also permitted evasion of important aspects of the problem. More particularly, the justified self-congratulation by the British military over relations with government looks increasingly problematic in terms of possible future political scenarios. In short, military discussion of operational capability rests on an implicit understanding of politics, both in terms of relations with the government and social politics, that may not be sustainable and that require attention. However, they are far less heeded in military planning than, for example, the role of NATO in the event of conflict in the Middle East.

Let us turn to a hypothetical example in order to illustrate the point. It is 2020. A food scare has led to a European government instruction for the destruction of all sheep older than six months. Combined with a serious economic downturn in agriculture, this has resulted in major acts of civil disobedience, especially the blockage of communication routes and tax strikes. The police are unable to deal with the situation, and unwilling to guarantee the safety of

European tax inspectors. English local government demands for hardship funds have been ignored, and the European Commission orders the army to open all major communication routes. As units deploy, officers and men reveal an unwillingness to use force.

Fanciful?[4] Similar disturbances occurred on a number of occasions in Western Europe in the 1990s, most prominently with strikes by French lorry drivers. Such problems were a world away from deployment in Bosnia, Kosovo or the Gulf, but the notion of a stable home base and of operations as a matter of out-of-area capability, the central focus in current British (and American) planning, is problematic. Given the relationships between police and local government, and the limited operational capability of police forces, it is likely to be upon armies, or indeed *the* army, that the European Commission relies. Possibly, the situation will, at least to a faint degree, echo that of the use of force to maintain cohesion in the Soviet bloc after 1945.

To turn from armies to air forces and navies is generally to downplay the issue of domestic order. For these services, it is easier to think in terms of a technologically driven model of military purpose and change, specifically today the Revolution in Military Affairs (RMA) supposedly offered by the military potential of advances in information technology. These advances are seen as providing an opportunity to conduct operations with more knowledge about opponents than ever before, and thus both to devise more finely tuned strategies and tactics and to execute them successfully. Information is designed to improve quick response capability and to transform tactics. For example, new weapons able to operate at night and to overcome camouflage and foliage are but part of a new ability to discern presence and movement.

Improved information makes it easier to conceive of manoeuvre rather than attritional warfare. Instead of a general sense of relative capability, it is possible to aim for an exact understanding of the opposing system and of its relations with one's own, and thus to aim for high tempo: a faster rate of activity that benefits from, and sustains, the advantage of manoeuvre. Thus, information provision made it possible to abandon the highly controlled style of warfare that had predominated in training for regular conflict between major Cold War forces: manoeuvre was to replace structured positional encounters. This also accorded with the greater professionalism and higher levels of training associated with forces abandoning conscription. Manoeuvrability was also linked to the weight of modern weaponry, the increased lethal quality of individual units, and the smaller size of modern forces. There was a counterpart in the emphasis on information and mobility in modern policing.

Aside from more information – for example, the use of satellite-linked global positioning systems to locate units and targets in real time – there would also be information warfare, for example cyberwar. Attacks on the information networks of rivals would have direct military value as part of 'deep battle', and would also threaten their political and economic networks.[5] Information technology played an increasing role in training. For example, Western forces came to employ imaging GIS (Geographical Information System) software in order to provide instantaneous two- and three-dimensional views of battlefield exercise areas.

Such training became more necessary as the combat experience of commanders and units declined after the 1914–45 peak, and also as increasingly sophisticated weaponry was introduced. Furthermore, the synergy of forces in different spheres – ground, sea and air – was crucial to the theory of the RMA, and this required both the development of new organizational structures and careful training for commanders and units. Belief in the interdependency of ground, sea, and air, and its potential reflected not only technological developments, but also the experience of the American campaign against Japan in the Second World War. This saw the most intensive rate of amphibious operations in history, as well as the most important example of air–sea conflict. The fate of British warships off Malaya in 1942 and of their Japanese counterparts in 1945 indicated that air power was not an optional add-on. If command of the sea meant anything, it had also to mean command of the air. Doctrine was translated into practice with, for example, the rapid development and deployment of the necessary landing craft for American and British amphibious operations.

A synergy of land and air – and, more specifically, of the different branches of land warfare – was also seen in Europe in the Second World War. The development of anti-tank weapons led to an emphasis on the co-operation of armour with infantry and artillery. Furthermore, the Allied powers, especially the Soviets, placed a heavy emphasis on artillery in their offensives against the Germans.[6] Thus, the course of the war established the clear superiority of the practice and doctrine of combined arms operations.

The RMA drew on the ideas of the US Army, specifically the Training and Doctrine Command high-tech concept of the Air Land Battle Doctrine which was promulgated in 1982 in order to offset Soviet numerical superiority on the Northern European Plain and, more generally, between the Baltic and Austria. Air Land was an addition to forward positional defence which, as planned for in this period, aimed to defend as much of West Germany as possible – a political requirement demanded by the German government. Air Land and the resulting military system was not tried out in conflict in Germany, but they nevertheless influenced notions of effective synergy between land and air and of a combination of firepower and manoeuvre.[7] Air Land also matched the Soviets in recognizing a level of conflict (and thus planning) in war between tactics and strategy. This was termed the operational level.

Advanced technology greatly contributed to ideas of Air Land battle. The Americans planned to employ stealthy attack aircraft and 'smart' guided weapons fired from stand-off platforms. Laser-guided projectiles and programmed cruise missiles would inflict heavy damage on Soviet armour, while advanced aircraft, such as the F-15 and F-16, would win air superiority and also attack Soviet ground forces. Stealth technology would permit the penetration of Soviet air defences, obliging the Soviet to retain more aircraft at home, and would also threaten their nuclear deterrent. The bombardment of Soviet forces would be enhanced by cruise missiles and by attack submarines firing missiles. Co-ordination would be made possible by computer networking, a new generation of spy satellites with six-inch resolution, other important sensors, such as the AWACS aircraft, and the Global Positioning System.

This was a strategy awaiting its place in military history. It also required a rethinking of established doctrine that had major implications for the training of commanders already in place. This has been an important aspect of late nineteenth- and twentieth-century command, and has led to a different type of military professionalism to that of earlier centuries.

In some respects the emphasis on operational synergy, and thus the multiplication of force, was, and is, essentially another version of long-established arguments for flexibility and for a focus on maintaining the tempo of attack. It would not have surprised Napoleon or Moltke, and can be found in the latter's *Instruction for Large Unit Commands* (1869). In some respects, this tradition, with its emphasis on victory in battle, is one of a tactics-led strategy that needs to be matched by a careful political exit plan. As such, it reflects the application of science to weaponry that has been such a prominent feature of the last century.

The technologically driven model of modern warfare, however, is of only partial value. It may help in the discussion of confrontation between symmetrical powers and between particular weapon systems, but does not address adequately the issue of asymmetry. Nor does it answer the question of how to turn victory into a successful settlement – the problem that thwarted Israel after 1967. The Israeli example is especially pertinent as, in successive wars, Israeli forces defeated large, more numerous Arab opponents, only to find peace elusive. Their manoeuvrist, high-tempo style of conflict, with its emphasis on armoured attacks, in which they had clear advantages over their opponents, was replaced by the positional defence of retaining control over occupied territory by the use of infantry, with the initiative resting with their assailants. The Israeli army suffered greatly from the absence of a co-operative local police force.

Here politics reappears. Why people fight and why they are willing to kill and to risk death are questions that go to the heart of military capability. In Britain, the issue is linked to the historical legitimacy of the military as the monopoly provider of organized force. This legitimacy is a conflation of a number of factors. The political history of the last eighty years is important. Although deployed in South Wales before the First World War and in Glasgow in 1919, the army was not used to suppress the General Strike of 1926 or the Miners' Strike of 1984–5, or to deal with autonomous movements in Scotland or Wales.

Furthermore, the British armed forces have had a 'glorious' past since 1943, certainly compared to some previous periods. The British benefited greatly from Allied assistance in the second half of the Second World War. Thereafter, there were no lengthy and unsuccessful wars of decolonialization comparable to Algeria or Indo-China for France, the Suez expedition of 1956 was cut short, and Harold Wilson's rejection of American pressure to help over Vietnam, as did Australia and New Zealand, was important to the continued prestige and depoliticization of the British armed forces. There was no equivalent to the involvement in politics that affected the French army or to failure in Algeria, or to the Vietnam experience for the Americans.[8]

In the British case, it is unclear how best to protect these advantages. The notion of defence co-operation with other European Union countries, especially France, that was much discussed in the 1990s and is currently on the table presumes a contractualism on the part of the British government – an ability to retain final control over identity and operations – that does not match the drive for pooled sovereignty that is currently being pressed.

In part, a problematic political context for the armed forces will be a return to the past. Yet for the British, as for other armed forces, there will be a different social politics. The nature of military society and service, including literacy, political awareness and discipline, in the seventeenth and eighteenth centuries was very different to the situation today. In the past, there were different requirements from soldiers and sailors, specifically far more of a need for physical strength and far less of an obligation to master complex processes that needed a long period of training.

As the nature of Western society became less deferential and more individualistic, hedonistic and even rebellious from the 1950s, so military society came to be very different to the rest of society, and to seem even more so. In reaction to the losses of war in 1914–45, and in accordance with new cultural moods, there was an abandonment of military triumphalism and a widespread feeling that victory was not worth the price. This did not lead to an automatic pacifism. For example, despite their losses in the First World War, the British proved willing to fight again in the Second World War. Some of those who saw Louis Milestone's markedly anti-war film *All Quiet on the Western Front* (MCA/Universal, 1930) were fighting across the same battlefields a decade later. However, the reaction against war became stronger from the 1950s and, more particularly, the 1960s. Again, this widened the gap between society and the military. From at least the First World War, Western governments had been acutely concerned to assess the attitudes of soldiery and civilians, not least towards the prospect of conflict and the course of war. The 'information state' registered both wartime disaffection and the long-term shift in attitudes towards military service.

The hierarchical character of military organizations is, indeed, very different to that of modern society. Given the role of the military as wielders of force, it is unlikely that such a hierarchical structure will be abandoned and that the self-managing teams increasingly seen in modern economies will be matched. On the other hand, modern armed forces seek to train ordinary soldiers and junior officers to take decisions on their own.

From the 1950s, social and military trends interacted to affect recruitment and training in Western societies. One specific problem related to training. As the aptitude and skills the military required rose markedly, so a greater percentage of the civilian population became unfitted for service, certainly for skilled service. This greatly affected the value of conscript forces, indeed ensuring that such forces came to have a lesser relative capability. Furthermore, the logistical and financial burden of maintaining conscript forces, not least in the context of rising living standards and expectations, was such that the maintenance of the force became the prime military function. This affected some Western European states in the 1990s.

For other states, including the USA and Britain, the actual and potential problems posed by changes in society and in military requirements were, *in part*, ended with the abolition of conscription. The military and political history of British policy towards Northern Ireland, for example, would have been very different had there been a conscript army. Conscripts might have been unwilling to serve, and the deployment and tactics employed might have placed a greater emphasis on avoiding casualties. However, despite the easier command culture created by the end of conscription, it is unclear how happily the reasonably well-educated volunteer British armed forces would take to being part of a European army.

The Northern Ireland problem is a prime instance of the difficulties that domestic divisions can cause for the external military capability of a state. It is also a reminder that democratic processes do not necessarily end or even limit such problems. Concern over the situation is not new. In 1807, Colonel Hawthorne wrote from Dublin:

> a divided or distracted people like us are not calculated to meet such an invader as [Napoleon] Bonaparte . . . the 12th of this month instead of lamenting over the fatal consequences of the battle of Friedland [a Napoleonic victory over Russia], the Orange Yeomanry of this kingdom were celebrating a battle [of the Boyne, 1690] fought upwards of 100 years ago, with every mark of triumph and exultation as if Ireland had no other enemy than its Catholic inhabitants.[9]

Whether modern forces are conscript or volunteer, they are still affected by other aspects of the RAM (Revolution in Attitudes to the Military). Aside from views within the military, on the part both of commanders and of other ranks, there is now within Western societies a widespread political unease about causing, let alone receiving, casualties. This focuses in particular on civilian casualties. Thus, in 1999, Serb civilian fatalities during NATO bombing operations in the Kosovo war became a major issue, far more so than such losses in previous conflicts. More generally, the same attitude can be glimpsed in concern about Western hostages during insurrectionary wars. In part, this is an aspect of a wider social shift within the West that can also be seen, for example, in opposition to the death penalty.

This shift also affects attitudes towards the death of combatants, both in one's own forces and in those of opposing forces. For example, there was American disquiet about losses in the Gulf War among retreating Iraqi army units in 1991, and controversy about the sinking of the Argentinian battleship the *General Belgrano* by a British submarine during the Falklands War in 1982.

As far as countries' own combatants were concerned, there was particular sensitivity on the part of American leaders, as seen, for example, during the Kosovo crisis of 1999. This was traced to public disquiet about what were minor losses in an earlier intervention in Somalia in 1992–3, OPERATION RESTORE HOPE. The Americans, originally 28,000 strong out of a 37,000-strong United Nations force, were down to about 4,500 troops from May 1993, but on 3 October a Somali ambush of a US Ranger force led to the death of 18 US soldiers, the wounding of 78 and the capture of one.[10] This was judged unacceptable. It was rapidly followed by the abandonment of aggressive operations by the US troops in Somalia. Furthermore, thereafter none were sent on peacekeeping missions to Africa.

Political and public reluctance to accept losses represents a shift from the militarization of American culture that was so pronounced in the 1940s and 1950s.[11] Concern about losses is also a marked shift from attitudes early in the twentieth century. For example, in 1905 a British Staff College conference was told that the experience of previous British operations in Afghanistan, and of British losses in the Boer War, suggested that in the event of another major conflict there it would be necessary to assume manpower losses of 20 per cent per quarter, 80 per cent per annum; and even then that this assumed battle losses less than those in the contemporaneous Russo–Japanese conflict in Manchuria.[12]

In contrast, in the USA the collapse of public support for the war in Vietnam in 1968 ushered in a period in which casualties ceased to be acceptable.[13] It was far from clear that this is true to the same extent of other Western powers, although none faced losses comparable to America's in Vietnam. Both France and Britain seemed, over the last quarter-century, to be willing to risk casualties, although it is less clear that this is also true of other Western European countries, such as the Netherlands or Germany. Casualties in both Northern Ireland and the Falklands War did not lead Britain to abandon either.

The media pressed the American military from 1968, scarring civil–military relations and revealing what is today termed the 'CNN factor': the need for military operations to deal with hostile and sensationalistic media coverage in an age of instant communication. More generally, the media has acted to demand intervention abroad, while at the same time posing serious operational constraints. Although prefigured with the rise of the mass daily press, the adult male franchise and widespread literacy in the late nineteenth century,[14] the media has become more insistent in a visual world. Furthermore, management of the media is a particular problem with coalition campaigns, as they bring a multiplicity of media organizations, although it is the American media that is most influential in its home market, and also most powerful on the global range.

Despite the popularity of war films and toys,[15] there has been a reluctance to serve in the military. Difficulties in recruitment and retention affected volunteer forces such as the British. They also affected conscript forces. In August 1999, Carlo Scognamiglio, the Italian Defence Minister, warned that Italy would be unable to guarantee its own security, let alone meet its foreign peacekeeping obligations, unless the number of professional soldiers was doubled from 25,000 to 50,000. He blamed Italy's falling birth rate and a liberal law on conscription allowing people to opt out of military service as conscientious objectors. In 1993, when 118,000 of the 146,000-strong army were conscripts, the Spanish government decided that it could not contribute a brigade to the UN forces in Bosnia, as conscripts could not be expected to serve and there were insufficient regulars. Most Spanish men of draft age never go into uniform, and Spain plans to abolish the draft by 2002. In the Gulf War, the capability of the French Force d'Action Rapide was affected by the inability to demand that the 40 per cent who were conscripts serve there. In 1996, the French government announced the abolition of conscription by 2002, and a cut in the planned size of the armed forces. Germany plans to keep the draft, but 40–45 per cent of those eligible opt for civilian service, for example as paramedics.

After the collapse of the Soviet Union, the Russian armed forces were badly affected by a crisis of manpower: exemptions from military service rose, as did conscription evasion, and desertion. The terrible conditions of service were well publicized, and public and political support for the notion of military service fell.[16] Aside from a reluctance to serve, in Western societies there is also an unwillingness to spend, shown in significant cuts of the percentage of national wealth and government revenues spent on the military: in 1999, in Germany this dropped to 1.5 per cent of GDP, leading to public complaints from the American Defense Secretary.

More generally, the RAM can be linked to changing notions of masculinity in Western society, specifically a downgrading of the view that martial values are important for men. The RAM was also linked to the rising prominence of women and female attitudes in Western societies, especially the Anglo-American world. In general, although women have played a major role in the social contexts of warfare throughout history,[17] women seem to be less sympathetic to bellicose values, and less willing to consider a resort to war. For example, a Gallup poll of the mid-1990s revealed that American women were far less sympathetic than American men to the use of the atomic bomb against Japan in 1945.[18] The demolition of national images and myths and the desecration of national symbols, such as the flag, were other aspects of a departure from traditional values.[19]

As far as the USA is concerned, the RAM, in part, involves withdrawal from risk, certainly risk understood in terms of suffering casualties. This has not, however, been accompanied by a policy of isolationism – indeed, far from it. Both the Cold War and, subsequently, the liberal interventionism called for in the 1990s to further a humanitarian world order involved far-flung commitments. There has thus been a gap – a demand for the deployment of effective military threats alongside an unwillingness to suffer casualties. Technology has been seen as the panacea for this gap, and, in part, the notion of the RMA is the consequence of the gap. There has to be a belief in technology in order to bridge the gap. The RMA also fulfils the cultural predisposition of the American military to believe in the possibility of the successful use of overwhelming force to achieve decisive victory, not the pursuit of limited goals by limited means and with limited resources.[20]

The RMA, in some respects, mirrors the earlier confidence that the atom bomb would give the USA a decisive advantage in any clash with the Soviet Union. This made it possible to demobilize Western troops after the Second World War, and thus to reduce the political and military costs of commitment to Western Europe and confrontation with the Soviet Union. However, the value of this strategy and weaponry was limited. First, even before the Soviets unexpectedly gained the atom bomb in 1949, it was clear that the weaponry was not sufficiently flexible (in terms of military and political application or acceptance of its use) to meet challenges other than that of full-scale war. Thus, the Americans did not use the atom bomb, of which they then had very few, to help their Nationalist Chinese allies against the Communists in the civil war of 1945–9. Similarly, it was not employed against the North Koreans. Instead, that war was fought with a strengthened conventional military. Second, the Soviet acquisition of the weaponry greatly lessened its potential value. The USA's efforts to leap ahead with the hydrogen bomb were swiftly thwarted by the Soviet Union's development of the same weapon.

The belief that technology can bridge the gap between the goals of effective reach and no casualties is unjustified. The essential fault is political, both the taking on of a series of major commitments and the encouragement of the belief that this can be problem-free. Military leaders have warned of the former, especially of over-stretch. For example, both American and British generals argued that commitments to first Bosnia and then also Kosovo spread their forces overly thin. However, these warnings have been neglected or simply treated as part of the lobbying process.

Military leaders also suffer from the application of commercial attitudes to planning: namely, a commodification of military resources so that they are akin to those of industrial parts or commercial products, and available just in time. As business experience shows, however, just in time can mean just too late. Furthermore, there is the danger that the military will become a 'hollow force', lacking adequate reserves to be properly capable of combat – in short, that it will be effective as a peacetime force, but not for war.

This demand-driven model also suffers from the unpredictability of the political demands placed on the military, from the very uncertainty of conflict itself, and sometimes from an underrating of the problems of transferring military resources from one type of conflict to another. These resources are less transferrable now than was the case with earlier military technologies and systems, with their heavier emphasis on the infantry, especially on soldiers able to carry their own supplies, and the lesser need for maintenance of complex equipment. This is a major example of the deficiencies arising from technological advances. They not only fail to live up to expectations, but also frequently represent a deterioration in the effectiveness of particular characteristics of the earlier system.

In 1999, Western policy over Kosovo presupposed no serious simultaneous deterioration in ongoing confrontations with Iraq and North Korea, as well as a military confrontation with the Serbs that did not require the occupation of Serbia or the suppression of large-scale guerrilla forces. In the event of optimistic scenarios in these and other crises proving misplaced, there is scant fallback in terms of additional resources. This is paradoxical, given the unprecedented wealth and borrowing capacity of major states, the size and sophistication of their economies, and the amount spent on armed forces. However – and here it is worth returning to the organizational theme of Chapters Two and Three – this expenditure has been spent on a steadily more specialized military. Indeed, the

RMA and other notions of the late twentieth century can be seen as the ideology of a very specialized and very high-cost military. The cost of advanced weapons has risen far faster than the rate of inflation since 1945.[21] The cost and difficulty of maintaining and moving units have also risen considerably. For those very reasons, this military cannot be used extensively, for the availability of additional resources that are already ready for military use is very limited. This limitation, 'blunting' or redundancy of the military is a counterpart to the restrictions created by shifts in the public acceptability of conflict and casualty and, more specifically, of the inhibiting effect of the destructive potential of nuclear weaponry,[22] although it is unclear, in the long term, how inhibiting this effect will be.

Furthermore, the inhibiting effect of this nuclear potential served as much to increase interest in defining a sphere for tactical nuclear weapons and in planning an effective strategic nuclear first strike as it did to lessen the chance of a great power war, or to increase the probability that such a conflict would be essentially conventional. During the Cold War, the crucial strategic zone was defined as the North European Plain, and the Soviet Union had a major superiority in conventional forces in this sphere.[23] The threat of retaliation by American nuclear strength, both tactical (e.g., one-mile range atomic bazookas) and strategic, served to lessen their threat, but in the 1970s, the Soviet Union was able to make major advances in comparative nuclear potency, producing a situation in which war was seen as likely to lead to MAD (mutual assured destruction).

This led the Americans both to press in the 1980s for the development and deployment of new nuclear and anti-missile weaponry, and also to develop ideas for countering Soviet conventional supremacy. In response to Soviet mass, and in place of the static forward defence that had been hitherto planned, there was an emphasis on manoeuvrability, or flexible response. This was given a high-technology resonance with the doctrine of Air Land Battle.

While strategists seek to wrestle with such problems, it is also necessary to consider how best military theories can be applied to less 'advanced' militaries, or whether, instead, a different literature, planning and training should be encouraged. The historical perspective offered in earlier chapters reveals the complexity of the notion of military progress, the continued viability of 'less advanced' militaries, the range of factors that led to a failure to adopt advances, and the general lack of theoretical and scholarly attention paid, both at the time and subsequently, to 'less advanced' practices and traditions. Despite globalization and the 'shrinking' of the world, this situation will continue. If, for example, Tudor muster rolls reveal that bowmen were still the mainstay of the late Elizabethan militia because the government was reluctant to face up to the cost implications of training and equipping musketeers, there are numerous modern equivalents.[24]

The convertibility of human, financial and economic resources into war is far lower in the West today than ever before. This involves serious military problems in any asymmetrical conflict with a society whose military is less sophisticated and where the rate and speed of convertibility are higher. In symmetrical confrontation or conflict, the smaller size of modern militaries will be matched in the case of likely opponents; but it is a very serious problem in the case of asymmetrical warfare, for example the occupation of territory in the face of a hostile civilian population capable of mounting effective opposition.

This can be glimpsed by comparing the very difficult German occupation of Serbia in 1941–4 with what might have happened in 1999 had a war over Kosovo led to the occupation of much of Serbia. Thanks to helicopters, it would have been easier to operate in Serbia in 1999, but it is difficult to feel confident that a long-term occupation would have been successful, especially if (as in 1941–4) resources had been stretched by commitments elsewhere in the world. In April 1941, the Germans lost only 151 men overrunning Yugoslavia in just over a week, but subsequent campaigning against partisans proved much more costly. Including allied (Italian, Croat and Bulgarian) forces, the Germans deployed over 120,000 men against the partisans in an encirclement operation in May 1943, while that September, ten German divisions alone were in use. On

1 January 1944, the German, Bulgarian and quisling force amounted to 360,000 men.[25] Such forces were simply not available in 1999.

The smaller number of units available in the militaries of advanced states, compared to the situation fifty years ago, has many implications for the way in which war can be fought. It ensures a lower density of units, and therefore enhances the possibility of manoeuvre and the need to adapt training and doctrine accordingly. Lower density affects both defence and attack. It is likely that a future sphere of conflict will be more like that of the Napoleonic period and less like that in the two World Wars, in that operations will be more fluid and less constrained by front lines. On the other hand, the secure movement of supplies is dependent on having areas that are under control. Manoeuvrability will be linked to an emphasis on gaining and retaining the initiative in order to dictate the battle (very different requirements from those required for peacekeeping). The quest for the initiative will put a premium on gaining and using information. The smaller number of units also reduces the availability of reinforcements and of forces capable of controlling conquered territory.

At the time of writing, major conflicts were under way in a number of countries, including Afghanistan, Angola and Congo/Zaire, and on the Eritrea–Ethiopia border. In many other countries, armies were involved in arduous and protracted counter-insurgency struggles, while in other areas, such as South Asia and the Middle East, forces prepared for imminent conflict with nearby states.

Allowing for the difficulty of assessing developments and the variety of factors playing a role in particular struggles, it is still worth probing for common or frequent features. The most apparent is the danger of assuming a Western or advanced model of military organization, and thus conflict. Outside the West, there is a range of military organization, from forces that are similar in many respects, not least in training and weaponry, for example those of India, Israel and Pakistan, to others, for example in Afghanistan, that do not measure up to modern Western conceptions.

To focus on the latter, it is the porosity of armed forces that emerges most clearly, in particular the overlap between such forces and civilian society. In place of a Western-style military professionalism, or a bureaucratic type of conscription, such as existed in the West for much of the twentieth century, there is a form of mass conflict, but one in which military service, although often compulsory, cannot be described as bureaucratic. Furthermore, those serving are not necessarily clearly distinguished from civilian society. Many do not, for example, wear uniforms. Yet this may be reconceptualized by arguing that, in these countries, civilian society, as understood in the West, does not really exist for adolescent and adult men. Instead, the defence of their community and the pursuit through violence or the threat of violence of particular goals is part of their life, just as other aspects of community life and their job are.

Given the mass character of military service in such societies, it is not surprising that armaments differ from those employed in advanced forces, although this is more true of the range of armaments than, for example, of the firepower of hand-held weapons. Furthermore, it would be mistaken to imagine that less 'advanced' forces have not benefited from aspects of what has been termed a silent revolution: namely, the marked development in conventional weaponry after 1945.[26] Still, these forces have no – or fewer and less sophisticated – planes, helicopters and ships, let alone satellite reconnaissance systems and long-range missiles. As far as land warfare is concerned, there is less mechanization and less specialization.

Yet as, for example, the Afghan conflict of 1979–89 amply demonstrated, the capacity of such forces to resist major powers is high. This is especially so if these forces receive foreign weaponry, but Afghan success against the British in the nineteenth century suggests that this factor should not be overrated. In the 1980s, the Afghan guerrillas received sophisticated American ground-to-air Stinger missiles, rocket launchers, mortars and radios. The Stingers reduced the effectiveness of Soviet helicopter gunships, and this helped to determine the outcome of fighting.[27] Yet, as so often, it is necessary to set the warfare in a wider context. The bellicose nature of Afghan society made it difficult to subjugate, the fissiparous character of its politics and the weakness of its central government structure made the country difficult to control, and the agrarian nature of its economy made it less vulnerable to military pressure. Furthermore, the Soviets did not appreciate the extent of the resistance they would encounter, or deploy their best forces.

Other examples of technologically more advanced powers encountering serious military problems included India, Israel and Russia. In Kashmir, the Indians found it easier to confront (and, in 1999, fight) similarly armed Pakistani regulars on the frontier than to tackle Muslim insurgents within Kashmir. Similarly, the Israelis were able to eliminate the Syrian missile sites in Lebanon when they invaded in 1982, but found it difficult to deal with popular resistance movements in south Lebanon and on the West Bank of the Jordan. In the latter case, the limitation of regulars in the face of popular resistance, of heavily armed men confronted with stone-throwing demonstrators, was crucial. In the former, the Israelis faced a guerrilla force, the Hizbullah, willing to take casualties, enjoying foreign support, and able to respond tactically to such Israeli advantages as air power, while still maintaining pressure.[28]

The Russians encountered considerable problems in the Caucasus, where Islamic independence movements were able to rely on considerable popular support, as well as on the terrain. The Russians invaded the region of Chechnya in December 1994, capturing the capital Grozny in 1995, but thereafter success in crushing resistance proved elusive, and eventually the Russians had to accept local autonomy. The 1994–6 war revealed the deficiencies of the badly led, poorly equipped, badly trained and motivated and under-strength Russian forces, not least in counter-insurgency warfare, although, given the number of Chechens, they did better than is generally appreciated.[29] The renewed Russian attack in 1999–2000 led to the fall of Grozny but revealed similar military deficiencies and a reliance on force rather than negotiation.

These examples could be multiplied if many of the struggles within Africa are considered. Thus, Ethiopian attempts to suppress secessionist and autonomist movements in Eritrea and Tigre, or long-standing Sudanese efforts to control the south have revealed the strengths of resisting forces,[30] as has, for example, the lengthy conflict between the government of Angola and its UNITA opponents. At times, the capability of the anti-government forces directly focuses on the limitations of advanced weaponry, at others on the ability to counter such weaponry with other weapons that do not require a sophisticated infrastructure. Thus, in Angola, the government's conventional forces find that their operational effectiveness declines in the wet season, which instead favours UNITA's guerrilla tactics. In 1999, the government sought to compensate with intensive attacks by Russian-made planes on UNITA bases. It is claimed that they have employed napalm and defoliants in addition to conventional explosions. The fighting on the ground is difficult to evaluate, not least because of deliberate misinformation. In the southern Sudan, the ground-to-air missiles of the Sudan People's Liberation Army have made aerial resupply, and air power in general, hazardous.

This is an instance of one of the major deficiencies with the emphasis on technological supremacy: namely, the degree to which a lead in weaponry is countered with new weaponry or with the development of relevant tactics and/or strategy. Thus, the development of aerial supply in the 1950s, and its enhancement from the 1960s by effective and more powerful helicopters, helped to overcome guerrilla challenges to land supply and communication routes. Thanks to helicopters, offensive operations could be mounted by helicopter, sometimes, as in Vietnam, in large-scale and sustained operations. However, aerial resupply and attack capabilities were in turn limited by anti-aircraft weaponry, limiting the safety of low-level operations and greatly affecting the vertical space of the aerial battlefield. In the 1990–1 Gulf War, the Iraqis employed mobile missile-launchers and mobile radars that 'locked' on to aeroplanes.

Civilian opposition and insurrectionary movements together, and even separately, make the land a more difficult military environment to control, and even to use as a base, than in the past. In the 1990s, this encouraged military theorists and planners in the West to put a greater emphasis on the sea as a sphere for manoeuvre and as a base area. In part, this reflected the enhanced possibilities of sea-based forces. In addition to basing aircraft on ships and ballistic missiles on submarines, there was development in sea-based guided tactical missiles. Cruise missiles were fired from ships or submarines by the Americans, and then also the British, in conflicts or confrontations with Iraq, Sudan and Serbia in the 1990s.

Aside from the sea as a platform for mounting bomb and missile attacks, there was greater emphasis on the potential for amphibious operations, whether in attack, reinforcement, or for the defence or withdrawal of interests and people. The 1992 US Navy-Marine Corps doctrine, *From the Sea*, focused on warfare waged from the sea. The 1998 British Strategic Defence Review confirmed a shift from largely static, defensive forces in Europe towards an expeditionary posture. Amphibious operations offered the seizure of the initiative, manoeuvrability and combination of different arms sought by modern strategists.[31] Their value has always been high because so much of the world's population and economic power is located on or close to littorals, and because so many states have coastlines and are therefore vulnerable.

Furthermore, an emphasis on amphibious capability is useful to the small number of states that have possessed it. Indeed, throughout the last half-millennium, far more states have been able to mount offensive operations by land than by sea. As a consequence, a sea-based strategy is of use to naval powers not only because they have the capability, but also because it enables them to emphasize their distinctiveness, and thus to gain an important psychological advantage. Amphibious attacks in the age of sail could be very successful (although against the colonies of other European powers, rather than against these states themselves), but their operational potential improved in that of steam.

The navalist emphasis on ship-to-ship engagement, however, overshadowed these operations. Winning dominance of the sea had to come before its use for force-projection – reasonably so – but this led to a neglect of planning for the latter. This relative neglect was accentuated by the greater prominence of commerce raiding and protection that arose from the development of the submarine, and from the relative unimportance of amphibious operations in the first four decades of the twentieth century, especially during the First World War and in the inter-war period. The course of the former did not really bear out the claim in Charles Callwell's *Military Operations and Maritime Preponderance: Their Relations and Interdependence* (1905) that 'there is an intimate connection between command of the sea and control of the shore'.[32] Aside from the serious failure, both operational and strategic, of the Dardenelles campaign in 1915, there was little mileage in the idea that naval power could make a material difference to operations in Flanders – a view pushed by Churchill as First Lord of the Admiralty. Aside from the serious naval problem (operating inshore against a protected coast), there was also the difficulty posed by the strength of German forces with the mobility offered by the railway. Naval operations in the Adriatic, Baltic and Black Sea made little difference to the course of the First World War on adjoining land masses.

Emphasis in the 1980s and 1990s on this aspect of sea power was a product not only of technological advances, but also of awareness of the geopolitical shifts produced by the end of colonial bases, and anyway, by the greater potential vulnerability of land bases. This limitation was underlined in October 1983 when lorries full of high explosive driven by guerrillas willing to give up their lives destroyed the American and French bases in Beirut. As a result, the American Marines sent to Lebanon in 1982 were withdrawn in 1984. In contrast, American naval units in the Mediterranean were able to dissuade possible threats, for example from the Libyan air force.

The relative safety of the sea maintains the contrast between military capability on land and at sea that has been one of the themes of this book. Irregular forces operated far less at sea. There are parts of the world where piracy is a major problem. The most serious is off South-East Asia, especially in the Straits of Malacca and the South China Sea and off the Philippines. The long coastline of the Philippines and the heavily indented nature of the east coast of Sumatra provide pirates with a multitude of possible anchorages. Furthermore, the light craft they use do not need to dock in major ports. There is also much piracy in the Caribbean, although against yachts rather than merchant shipping.

Yet such vessels are not able to threaten the warships of major powers, and nor indeed do they seek to do so, because the economic rationale of piracy depends on avoiding such confrontation. Although modern warships are soft-skinned compared to their predecessors during the World Wars, their protection is enhanced by sophisticated surveillance systems linked, in particular, to anti-aircraft and anti-missile weaponry, and their range extended by aircraft, helicopters and missiles. At sea, it is easier to distinguish and assess other vessels, and thus to avoid the situation on land in which guerrillas can be indistinguishable from the civilian population. The sea, in short, can be known. As a consequence of the range provided by aircraft, helicopters and missiles, ships are far better able to mount attacks on hostile powers than their predecessors. In addition, the 'transition costs' of amphibious operations have been greatly lessened as the point of contact (and emphasis) of sea-based attacks has ceased to be simply the coastal landing zone: the littoral has replaced the coastline. Naval power offers reach, mobility and logistical independence. It provides a dynamic quality that was lacking from fixed overseas garrisons. It is also a military substitute for colonial possessions.

This emphasis on the sea is one that is only possible for a small number of powers, however. Since 1945, the Americans have dominated the world's oceans to a degree unmatched in history. Technology combined with economic resources to give their fleets a strength and capability that the British navy had lacked at the height of its power.[33] This position was challenged in the Cold War, both by the growth of rival navies and by developments in anti-ship weaponry, specifically missiles. Thus, in 1967, the destroyer *Elat*, the Israeli flagship, was sunk by Soviet missiles used by the Egyptians. The build-up of the Soviet fleet that began in the 1950s quickly made the USSR the world's number two naval power, a position eased by the wartime destruction of the Japanese navy and the post war decline of the British navy. No other European navy rose to comparable position. By the 1980s, the Chinese ranked third[34] and the Japanese fourth among the world's navies. Nevertheless, the building of new ships and the upgrading of weaponry and communications kept the Americans in the leading position. Thereafter, the principal change was that the Soviet navy became increasingly obsolescent in the 1980s and 1990s, as it proved impossible to sustain the cost of new units. Furthermore, in the 1990s, the Soviet navy was affected by the break-up of the Soviet Union.

During the twentieth century as a whole, naval power became even more of a high-cost exercise, and one that was driven by technological imperatives as, from early in the century, the sea was contested by first submarines and then air power. It became necessary both to devise countermeasures and to ensure that the various forces that could operate on, over and under the sea co-operated effectively. The need for this was amply demonstrated during the Falklands War in 1982: supported by submarines and aircraft, the British navy saw off the challenge from Argentinian warships and aircraft. Naval conflict in this war was far more important than it had been in, for example, the Arab–Israeli wars of 1967 and 1973 or the India–Pakistan conflicts of 1965 and 1971, all of which were waged between contiguous states.

Thereafter, the sea has essentially been used for sea-to-land operations rather than contested between marine powers. The limited number of states with any major naval effectiveness ensures that this is likely to remain the case. Nevertheless, as there has been an arms race in East and South-East Asia, and naval shipbuilding has developed in newly industrialized countries such as Argentina, Brazil and India, so there are indeed enhanced possibilities of naval conflict.

Although naval support was important, the wars of the 1990s were land conflicts. They were analysed by commentators in very different ways. The 1990–1 Gulf War was seen as a triumph for technology, and helped to resurrect popular and air force confidence in the impact of air power[35] following its apparent failure in Vietnam. The Gulf War tested the Air Land Battle and came at the moment when the American RMA was coming together. The

Americans used satellite surveillance, Patriot anti-missile missiles against Iraqi missiles, Tomahawk cruise missiles to bombard Iraq precisely, and guided bombs to the same end. B-2 Stealth bombers able to minimize radar detection bombed Baghdad – one of the most heavily defended cities in the world – and did so with total impunity and considerable precision. Thermal-imaging laser-designation systems were employed to guide the bombs to their target. Cruise missiles made use of the precise prior mapping of target and traverse in order to follow predetermined courses to targets that were actualized for the weapon as grid references. They used digital terrain models of the intended flight path, in order to provide precise long-distance firepower. Allied tanks successfully employed precise positioning devices interacting with US satellites in a Global Positioning System. Tanks also made effective use of thermal imaging sights.

Satellite imagery was also responsible for the rapid production of photo-maps. Near-real-time information and communication provided many more opportunities for individual units on the ground to take decisions. Other high-tech devices included unmanned air vehicles, electronic countermeasures, and fuel-air explosions. Very complex control systems were employed to maintain the pace of the attack on Iraq, particularly the air attack.[36] Compared to earlier conflicts, target acquisition and accuracy were effective, although the nature of the target and terrain greatly assisted. The Iraqis were defeated with heavy casualties, while their opponents lost very few men.

The result of the war, however, was not simply a consequence of technology. The Stealth bombers and fighters, Cruise and Patriot missiles, and laser-guided bombs did less well than was claimed at the time. In particular, their much-praised accuracy was less manifest in combat conditions than had been anticipated, especially that of the Patriots, which had a crucial role in the missile-to-missile war. In addition, British runway-cratering bombs were less effective than had been envisaged.

Command and control proved unequal to the fast tempo of the Gulf War; indeed, the system buckled under the combined strain of its very complexity and of this tempo. On the ground, supporting artillery fire suffered from deficiencies, including a lack of adequate ammunition, and poor integration at unit level. Air Land Battle proved more difficult in practice than in theory, not least due to the problems of synchronizing air and land forces under fast-moving combat conditions.[37]

Much of the success of the American-led Allied coalition in the Gulf War was due not to respective weaponry but to Allied, principally American, fighting quality, unit cohesion, leadership and planning, and Iraqi limitations in all four and in other respects.[38] The Iraqis surrendered mobility and the initiative by entrenching themselves to protect their conquest of Kuwait. The Allied task and the Iraqi response were suited to the American 'systematic, production-line approach to warfare',[39] and failed to encourage attention to the limitations of the latter. The Americans provided two-thirds of the ground forces and even more of the air power, and the strength of their commitment compensated for the operational and/or political limitations of some other members. For example, no Arab coalition forces sent troops into Iraq.

In the aftermath of the Gulf War, the Americans intervened in Somalia from December 1992. Initial success was followed by an eventual failure that, in part, reflected the failure of gunships to dominate and control the situation on the ground.[40]

Problems with advanced weaponry were extensively discussed during the Kosovo war of 1999, a conflict that was fought out in the full glare of attention and amidst considerable controversy about the effectiveness and potential of particular weapon systems. NATO claims about the destructiveness of air power proved greatly misleading. Instead, the Serbs, employing simple and inexpensive camouflage techniques, succeeded in preserving most of their equipment. Their eventual retreat seems to have been due to Russian pressure, to the maintenance of NATO cohesion, and to the threat of a NATO land attack, rather than to the lengthy air offensive which involved 10,000 strike sorties. It is, of course, possible to speculate as to the likely effectiveness of the land attack, had one been mounted. It was certainly unlikely to have bridged the gap referred to earlier, because NATO casualties would probably have been far higher than Allied losses during the Gulf War.

Most conflicts in the 1990s have been, and are, far more 'low-tech' than those just discussed, although, with regard to the combatant societies, the forces may be operating at their technological optimum. Whatever the technology, the prime military problem remains that of securing a situation in which a solution can be imposed and will be accepted; and this scenario continues to face military as well as political challenges. Consider two conflicts that began to receive international attention in 1999: the conflict in the Caprivi strip of Namibia and that in Dagestan in the Caucasus. Both were conflicts in which one side sought to overturn imperial territorial settlements, and each reflected the failure of states to incorporate minorities. In the former case, a secession attempt was mounted by the Lozi-speaking people of Caprivi, who are dominated by the Ovambo majority of Namibia. About 200 fighters of the Caprivi Liberation Army attacked key points in the province's capital, Katima Mulilo, in August. However, the ill-planned attack was swiftly defeated. This 'military' response was followed by one that is harder to categorize. Paramilitaries from the Special Field Force and members of the Central Intelligence Service seized large numbers of suspects and treated them brutally.

In Dagestan, Islamic rebels declared independence in August 1999. Initially, about 600 members of the Wahhabi sect, some of them Dagestanis, but many not, sought both to destroy Russian power and also to coerce the population of Dagestan, most of whom were moderate Muslims. Control of the villages was contested, as the rebels were not strong enough to seize the capital, Makhachkala. However, as in other struggles, rival forces essentially operated in different environments – the rebels on paths, the mechanized Russian forces on roads.

Similar conflicts occurred throughout much of the world and at the start of the new millennium. Welcomed as progressive, self-determination, like nationalism and decolonization, was also a cause of domestic and international instability. The principle of self-determination failed to address the issue of who was allowed to seek it. In 1960, the United Nations stated that all 'peoples' had the right to self-determination,[41] but it was, and is, not clear how 'peoples' were to be defined.[42] An ethnic notion of nationalism and anxiety about the loyalties of those not comprehended in this definition led to the brutality of ethnic cleansing, as, for example, after the First World War with the expulsion of Greeks from Turkey, and again throughout Eastern Europe after the Second World War, and in Rwanda and the Balkans in the 1990s.

Again a note of variety has to be struck, as conflicts arising from attempts to overturn imperial territorial settlements, specifically from minority aspirations and/or the policies of ethnically-defined governments, are more present in some continents than others. They are more insistent, for example, in Africa[43] and Asia than in the Americas. Furthermore, the ability to settle disputes short of conflict, especially of long-term warfare, varies.

The wars that broke out in the 1980s and 1990s were part of a general pattern since 1945 in which the incidence of inter-state wars declined, but that of intra-state conflicts rose. The increase in the number of independent states has not led to a diminution in separatist tendencies, although that is not the sole cause of intra-state conflict. Whatever the cause, such conflict tends to see more political direction than that in inter-state warfare, although such direction is also important in the latter.

The definition of war as conflict between sovereign states remains of only limited value, although it dominates military theory. Instead, it is necessary to devote attention to attempts either to challenge control of the state or to split from the state. In some cases, war may be little different from organized crime or from riots. Indeed, terrorist and insurrectionary groups turn to crime in order to finance their activities. The Shining Path movement in Peru is an extraordinarily destructive and brutal rebellious movement founded in 1990 that, in part, supports itself thus, although, in its social revolutionary terms, this is a case of fair taxation.

Similarly, there may be a continuation between religious radicalism and conflict. Efforts to define war in terms of international law, and thus to separate it from rebellions, piracy, booty raids, marauding and the like, are at best an extrapolation of conditions in part of the world that are not valid elsewhere, and cannot therefore delimit the role of the military.

This observation links the present with past situations outlined in earlier chapters. It is also appropriate to consider how far other themes can be reprised. That of counterfactualism is obviously pertinent. Had Iraq or Serbia possessed the atomic bomb in 1990–1 or 1999 respectively, then the response to the Gulf and Kosovo crises might well have been very different. Indeed, the diffusion of nuclear and other highly destructive weaponry became a more pressing issue in the late 1990s. Both India and Pakistan publicly tested nuclear weapons, and – as also did other states, such as North Korea and Iran – developed long-distance missiles capable of carrying such weapons. The pharmacology of terror greatly expanded with developments in bacteriological and chemical weaponry. These were not restricted to preparations for state-to-state conflict. Instead, in the 1990s, Iraq employed such weapons against Kurdish insurgents, as well as providing mustard gas which the Sudan government used against another long-standing insurgency, that in the south of the country.

The use of sarin nerve gas by Aum Shinrikyo, a Japanese sect, in an attack on the Tokyo underground in 1995 showed that such substances could be made and used by non-state organizations. The limited effectiveness of the sect's attacks, despite the considerable resources at their disposal, was less striking than the attempt to widen the terrorist repertoire. As with firearms and earlier weapons, state monopolization of the means of violence proved to be limited. This was underlined by discussion about how a crude nuclear bomb could be fairly readily manufactured.

Such a situation did not mean that all weaponry could be easily replicated, or that it could be replicated in the quantities deployed by major powers. However, the potential capability of non-state players is a reminder of the danger of assessing and planning for conflict simply in state-to-state terms, and the greater volatility of many of these players underlines the point.

Counterfactuals are readily present in state-to-state conflicts. Some of the conflicts of recent years, such as the India–Pakistan clash in 1999, were not unexpected. They grew out of unresolved issues, a practice of bellicosity on the part of at least one party, and military confrontation. The same would have been true had war broken out between North and South Korea, as, indeed, seemed possible in 1999. Yet, other recent wars were unexpected. This was true, for example, of the Falklands War of 1982 and the Gulf War of 1990–1. More generally, the entire process by which the Cold War came to a close with the fall of Soviet power was unexpected, and it is unclear that it had to lead to the conflicts that broke out in the successor republics and the former Yugoslavia.[44]

These counterfactuals are pertinent because they affect the military planning that helps determine capability: if armed forces are maintained on a fit-for-purpose basis, it is necessary to consider what purpose may be appropriate. For the states involved in the Cold War, there was no long-term decline in the perceived danger of conflict. Indeed, in the 1980s, there was a build-up both in tension and in the deployment or planning of new military systems. The Americans deployed cruise and Patriot missiles, and allocated funds for research into space-mounted weaponry designed, in particular, to use lasers in order to destroy both other satellites and missiles. Concern about increased and developing Chinese, North Korean, Iranian, Iraqi, Libyan and other long-range missile capability, including the growth of Chinese and Iranian submarine forces, and the test-firing by North Korea in August 1998 of a Taepodong 1 rocket over Japan into the Pacific led, in the late 1990s, to a renewal in American interest in a theatre missile defence (i.e., an anti-missile defence) system, and to discussion in Britain, Japan, Taiwan and other countries about the need for a similar system. Equally, the potential of satellite communication, surveillance and control systems, in which the Americans had a major lead, led to the development of anti-satellite weaponry, tactics and doctrine.[45]

In many respects, this was a world away from counter-insurgency warfare, although some aspects of advanced weaponry, such as satellite surveillance, were useful for the latter. Irrespective of these uses, the attitude of mind that tended to condition planning for war between nuclear powers was not one that readily lent itself to counter-insurgency conflict. This was accentuated by the institutional interests of the individual armed services.

To conclude, planning for total war between the major powers is now, despite the weaponry available, less pertinent than at any time for over a century. This reflects the dominant role of the USA within the West, the degree to which other Western states cannot wage war without American consent and co-operation (as the British discovered, negatively, over Suez in 1956 and, positively, in the Falklands War in 1982), and the decline of Soviet power. Aside from the USA, only China is able to act as a threatening great power. China's challenge is restricted by the circumspection of its military projection: there are American forces in East Asia, not Chinese forces in the West Indies. Since 1950, China has clashed with the Americans in the Korean War, and with India, North Vietnam and the Soviet Union, but all bar the first were short-term conflicts, and the Chinese were wary of military, and indeed diplomatic commitments further afield. Despite this, Chinese interests in East Asia do threaten war. Attempts to coerce Taiwan and the pursuit of territorial claims in the South China Sea threaten American allies, and there is also a powerful distrust between the two states that reflects their very different political cultures. The prospect of such a conflict will be probed in the next chapter.

All other possible major wars at present oppose the great power – the USA – to lesser states, or threaten conflict between the latter. In the former case, the possibilities and limitations of technology and the role of changing attitudes towards the use of force (RMA and RAM) are both pertinent, but so, also, is the more general issue of the difficulty of imposing a verdict in a complex world. This problem unites great-power and small-scale conflicts, and the present with the past – and, probably, the future.

NOTES

1. Report of a Conference and Staff Ride, as carried out at the Staff College, January 1905, PRO WO 33/2747, p. 37.
2. L.G. Schwoerer, *No Standing Armies: The Anti-army Ideology in Seventeenth-century England* (Baltimore, MD, 1984).
3. H. Strachan, *The Politics of the British Army* (Oxford, 1997).
4. A very different scenario emphasizing external challenges is offered by M. Duffy, T. Farrell and G. Sloan (eds), *European Defence in 2020* (Exeter, 1998).
5. See, for example, C.S. Gray, *The American Revolution in Military Affairs: An Interim Assessment* (Camberley, 1997); A. Krepinevich, 'Cavalry to Computer', *The National Interest*, 37 (1994), pp. 30–42; G.R. Sullivan and J.M. Dubik, *Land Warfare in the 21st Century* (Carlisle Barracks, PA, 1993); Sullivan and A.M. Coroalles, *The Army in the Information Age* (Carlisle Barracks, PA, 1995); A. Irvin, 'The Buffalo Thorn: The Nature of the Future Battlefield' *Journal of Strategic Studies*, 19 (1996), pp. 238–40, 245–6; L.W. Grau and T.L. Thomas, 'A Russian View of Future War: Theory and Direction', *Journal of Slavic Military Studies*, 9 (1996), pp. 508–12. Triumphalist views are offered in J.S. Nye and W.A. Owens, 'America's Information Edge', *Foreign Affairs*, 75, 2 (1996), pp. 20–36, and J.R. Blaker, *Understanding the Revolution in Military Affairs: A Guide to America's 21st Century Defence* (Washington, DC, 1997).
6. G. Niepold, *Battle for White Russia: The Destruction of Army Group Centre, June 1944* (1987).
7. J.L. Romjue, *From Active Defense to Air Land Battle: The Development of Army Doctrine, 1973–1982* (Fort Monroe, VA, 1984); A. and H. Toffler, *War and Anti-War: Survival at the Dawn of the 21st Century* (1993).
8. A. Horne, *The French Army and Politics 1870–1970* (Basingstoke, 1984).
9. Col Hawthorne to Henry Addington, 17 July 1807, Exeter, Devon Record Office, 152 M/C 1807/018.
10. T. Farrell, 'Sliding Into War: The Somalia Imbroglio and US Army Peace Operations Doctrine', *International Peacekeeping*, 2 (1995), pp. 195–213; M. Bowden, *Black Hawk Down: A Story of Modern War* (New York, 1999).
11. M.S. Sherry, *In the Shadow of War: The United States Since the 1930s* (New Haven, CT, 1995).
12. Conference Report, PRO WO 33/2747, p. 110.
13. For the argument that the media was not responsible for sapping popular support, see C.A. Thayer, 'Vietnam: a Critical Analysis', in P.R. Young (ed.), *Defence and the Media in Time of Limited War* (1992), pp. 89–115.
14. S. Badsey, *Modern Military Operations and the Media* (Camberley, 1994).
15. P.M. Regan, 'War Toys, War Movies and the Militarisation of the United States, 1900–85', *Journal of Peace Research*, 31 (1994), pp. 45–58; J.W. Gibson, *Warrior Dreams: Paramilitary Culture in Post-Vietnam America* (New York, 1994).

16. M. Orr, 'The Russian Armed Forces as a factor in Regional Stability', in C. Dick and A. Aldis (eds), *Central and Eastern Europe: Problems and Prospects* (Camberley, 1998), pp. 103–4.

17. L.G. De Pauw, *Battle Cries and Lullabies: Women in War from Prehistory to the Present* (Norman, OK, 1998).

18. B.J. Bernstein, 'Truman and the A-bomb: Targetting Noncombatants, Using the Bomb, and His Defending the "Decision"', *JMH*, 62 (1998), pp. 568–9.

19. R.J. Goldstein, *Burning the Flag: The Great 1989–90 American Flag Desecration Controversy* (Kent State, OH, 1996).

20. F.G. Hoffman, *Decisive Force: The New American Way of War* (Westport, CT, 1996).

21. F. Spinney, *Defense Facts of Life* (Boulder, CO, 1986).

22. E. Luard, *The Blunted Sword: The Erosion of Military Power in Modern World Politics* (1988); M. van Creveld, *Nuclear Proliferation and the Future of Conflict* (New York, 1993); S. Aronson, *The Politics and Strategy of Nuclear Weapons in the Middle East* (Albany, NY, 1992).

23. For NATO weaknesses, see J.A. English, *Marching Through Chaos: The Descent of Armies in Theory and Practice* (Westport, CT, 1996), pp 157–164.

24. A.J. King (ed.), *Muster Books for North and East Hertfordshire 1580–1605* (Hertford, 1996); G. Phillips, 'Longbow and Hackbutt. Weapons Technology and Technology Transfer in Early Modern England', *Technology and Culture*, 40 (1999), pp. 576–93.

25. R.M. Kennedy, *German Anti-Guerrilla Operations in the Balkans 1941–1944* (Washington, DC, 1954); R. d'Arcy Ryan, *The Guerrilla War in Yugoslavia, 1941–45* (Camberley, 1994).

26. G. Hartcup, *The Silent Revolution: Development of Conventional Weapons 1945–85* (Oxford, 1993).

27. M. Galeotti, *Afghanistan: The Soviet Union's Last War* (1995). For the early use of helicopters, C.R. Shrader, *The First Helicopter War: Logistics and Mobility in Algeria, 1954–1962* (Westport, Conn., 1999)

28. Z. Schiff and E. Ya'ari, *Israel's Lebanon War* (1984); A. Bregman, *War and Israeli Society* (2000).

29. A. Raevsky, 'Russian Military Performance in Chechnya: An Initial Evaluation', *Journal of Slavic Military Studies*, 8 (1995), pp. 681–90; A. Lieven, *Flaying the Bear: Chechnya and the Collapse of Russian Power* (New Haven, CT, 1998).

30. R. Iyob, *The Eritrean Struggle for Independence: Domination, Resistance, Nationalism, 1941–1993* (Cambridge, 1995); M.W. Daly and A.A. Sikainga (eds), *Civil War in the Sudan* (1993).

31. C.S. Gray, *The Leverage of Sea Power: The Strategic Advantage of Navies in War* (New York, 1992); G. Till, *Seapower Theory and Practice* (1994); T.L. Gatchel, *At the Water's Edge: Defending against the Modern Amphibious Assault* (Annapolis, MD, 1996); C. Parry, 'Maritime Manoeuvre and Joint Operations

to 2020', in M. Duffy, T. Farrell and G. Sloan (eds), *European Defence in 2020* (Exeter, 1998), pp. 45–63; E. Grove and P. Hore (eds), *Dimensions of Sea Power: Strategic Choice in the Modern World* (Hull, 1998); *British Maritime Doctrine* (2nd edn, 1999).

32. R. Smith, *The Requirement for the United Nations to Develop an Internationally Recognized Doctrine for the Use of Force in Intra-state Conflict* (Camberley, 1994), p. 21.

33. D.A. Rosenberg, 'American Naval Strategy in the Era of the Third World War: An Inquiry into the Structure and Process of General War at Sea, 1945–90', in N.A.M. Rodger (ed.), *Naval Power in the Twentieth Century* (1996), pp. 242–54.

34. D.G. Muller, *China as a Maritime Power* (Boulder, CO, 1983); J.W. Lewis and X. Litai, *China's Strategic Seapower: The Politics of Force Modernization in the Nuclear Age* (Stanford, CA, 1995). For China's leading opponent, S.B. Weeks and C.A. Meconis, *The Armed Forces of the USA in the Asia–Pacific Region* (1991), pp. 122–56 for naval and marine forces.

35. J.F. Dunnigan and A. Bay, *From Shield to Storm: High-Tech Weapons, Military Strategy and Coalition Warfare in the Persian Gulf* (1992); R.P. Hallion, *Storm Over Iraq: Air Power and the Gulf War* (Washington, DC, 1992). More generally on the conflict, see L. Freedman and E. Karsh, *The Gulf Conflict, 1990–1991: Diplomacy and War in the New World Order* (Princeton, NJ, 1992) and M.R. Gordon and B.E. Trainor, *The Generals' War: The Inside Story of the Conflict in the Gulf* (Boston, MA, 1995).

36. M. Mandeles, T.G. Hone and S.S. Terry, *Managing 'Command and Control' in the Persian Gulf War* (Westport, CT, 1996).

37. R.R. Leonhard, *The Art of Maneuver: Maneuver-warfare Theory and AirLand Battle* (Novato, CA, 1991), pp. 261–99; A. Bin, R. Hill and A. Jones, *Desert Storm: A Forgotten War* (Westport, CT, 1998).

38. R.H. Scales, *Certain Victory: The US Army in the Gulf War* (Fort Leavenworth, KS, 1993); R.M. Swain, *'Lucky War': Third Army in Desert Storm* (Leavenworth, KS, 1994); N. De Atkine, 'Why Arab Armies Lose Wars', *Middle East Quarterly*, 4 (1999).

39. R.M. Connaughton, *Swords and Ploughshares: Coalition Operations, the Nature of Future Conflict and the United Nations* (Camberley, 1993), p. 19.

40. R. Smith, *The Requirement for the United Nations to Develop an Internationally Recognized Doctrine for the Use of Force in Intra-state Conflict* (Camberley, 1994), p. 21.

41. General Assembly Resolution 1514 of 14 December 1960.

42. S.J. Anaya, 'The Capacity of International Law to Advance Ethnic or Nationality Rights Claims', *Human Rights Quarterly*, 13 (1991), pp. 403–11, and J.J. Corntassel and T.H. Primeau, 'Indigenous "Sovereignty" and International Law: Revised Strategies for Pursuing "Self-Determination"', *Human Rights Quarterly*, 17 (1995), pp. 140–56.

43. A. Clayton, *Factions, Foreigners and Fantasies: The Civil War in Liberia* (Sandhurst, 1995), and *Frontiersmen: Warfare in Africa since 1950* (1999); R.W. Copson, *Africa's Wars and Prospects for Peace* (Armon, NY, 1994); J.W. Harbenson, D. Rothchild and N. Chazan (eds), *Civil Society and the State in Africa* (1994); C. Clapham (ed.), *African Guerrillas* (Bloomington, IN, 1998); T.M. Ali and R.O. Matthews, *Civil Wars in Africa: Roots and Resolution* (Montreal, 1999).

44. J. Gow, 'After the Flood: Literature on the Context, Cause and Course of the Yugoslav War – Reflections and Refractions', *Slavonic and East European Review*, 85 (1997), pp. 446–84.

45. Institute for National Strategic Studies, *Strategic Assessment 1999: Priorities for a Turbulent World* (Washington, DC, 1999), pp. 304–5; J.T. Richelson, *America's Space Sentinels: DSP [Defence Satellite Program] Satellites and National Security* (Lawrence, KS, 1999).

TEN

War in the Future

The nature of conflict in the future is, of course, unknown, but it is nevertheless something that is planned for and thus subject to discussion. Both the development of nuclear weaponry and, subsequently, the end of the Cold War encouraged claims that war had become obsolete.[1] These were misplaced, not least because it is possible to fight without employing weapons of mass destruction and because the Cold War superpowers were not responsible for the conflicts of the states they armed.[2] Furthermore, the nature of the development of human society will doubtless impose burdens and encourage pressures both within and between states, posing problems for their governmental and social structures, for inter-state relations, and for conventions and institutions seeking to encourage regional and international peace. Historical experience suggests that years of peace are all inter-war as well as postwar periods.

There is certainly no shortage of issues to exacerbate international and intra-state relations. Both ideology and resources remain core problems, and the latter problem has widened to ensure that state attention increasingly focuses on the consequences of international developments such as demographic growth, large-scale migration, the spread of disease, and pollution. Ideology focuses a range of sometimes linked issues, including the practice of authoritarianism, not least at the expense of human rights, the relationship between states and separatist movements, and disputes arising from assertive

Islamic movements. Ideological and resource issues are inter-related, not least with the high population growth rates in North Africa and the Middle East, the presence of most of the world's known and readily exploitable oil reserves in the region, and the shortage of water there.

The massive global increase in population that took place in the twentieth century will continue for at least several decades, despite falling birth rates in many countries. The rise will reflect not only the entry into fertility of current children, but also improvements in public health and medical care, so that average life expectancy increases at all ages. A 1999 United Nations Population Fund Study suggested that the world's population had risen from 3 billion in 1910 to 6 billion in 1999, and would increase to 8.9 billion by 2050, although, thereafter, growth is expected to diminish. By the late 1990s, 78 million babies were born each year.

This rise in population has tremendous resource implications, and this will remain the case even if growth rates decrease. Aside from the rise in overall demand for employment and goods, there will also be a continuation in the rise in per capita demand. The move of much of the world's population into urban areas and their willingness to reject parental living standards and aspirations increase exposure to consumerist pressures. These pressures are also increasingly present in rural areas, not least due to the massive extension of access to television in countries such as India, and the role of advertising.

Demands for goods and opportunities will increase the volatility of many states, especially those that cannot ensure high growth rates and/or the widespread distribution of the benefits of growth. This will exacerbate problems of political management within and between states, encouraging the politics of grievance and redistribution. Conflict causes poverty, but poverty also encourages conflict. By 1999, 95 per cent of the rise in the world's population was occurring in developing countries, in many of which much of the population lacked adequate housing, sanitation and health services. Furthermore, in 1999, it was estimated that nearly a billion people were illiterate. This again increases volatility. Whereas in the

West, democratization may make it harder to persuade public opinion to support war, elsewhere popular pressures may have a different consequence.

Rising global demands will interact with a world in which the availability of resources and population pressures both vary greatly.[3] Although the most sensible ways to maintain and enhance resources require international co-operation, it is likely that confrontation and conflict will arise from unilateral attempts to redistribute resources. The situation is likely to be most acute with resources that can flow or move across borders, for example water, oil and fish. The limited supply and, in places, near exhaustion of such resources poses a major problem. Relations between Turkey, Syria and Iraq are strained by disputes over the rivers Tigris and Euphrates, and between Israel, Syria, Palestine and Jordan likewise over the river Jordan. Oil is also crucial to the course of a number of conflicts, providing revenues for governments such as that of Angola, and thus targets for opposing forces. In 1999, the Sudan People's Liberation Army attacked the government's oil pipeline at Atbara, demonstrating the state's vulnerability. Furthermore, the offshore availability of oil and fish ensures that these disputes will interact with quarrels about borders and territorial waters.[4]

Defence and security will focus on preserving access to resources, as states struggle to meet the needs of their populations and to defend their places in the global economic system. The 1999 Global Environment Outlook report, *GEO 2000*, produced by the UN Environment Programme, predicted environmental degradation and population growth, leading to the possibility, in the first quarter of the twenty-first century, of 'water wars' over scarce resources in North Africa and South-West Asia. Conflict is likely to encourage further environmental degradation, deliberately, as with the Iraqi destruction of Kuwaiti oil installations in 1991, or as a consequence of bombing for other ends, as with Allied destruction of Serbian petrochemical plant in 1999, and less deliberately. Although there has been talk of employing advanced weaponry without causing such degradation, it is difficult to feel confident on this issue.

To a certain extent, confrontation over resources will be a new version of the wars launched by revisionist powers (those seeking to change the status quo).[5] However, the wars are likely to be more acute, because resource issues will possibly make it easier to elicit popular support, and also harder to secure compromise. Such conflict will be civil and international. The former will provide the lightning rod for ethnic, class and religious tensions, and is likely to be particularly brutal. This will be in keeping with a general tendency of 'primitive' warfare to cause higher casualty rates than the majority of conflicts involving regular forces.[6] A recent example is the very high death rate in civil conflict in Rwanda, especially the Hutu-directed genocide of 1994, where many of the victims were beaten to death. The prospect of chaos is also apparent within states that have an advanced military. For example, in India there is a major contrast between the forces of the central government, which include nuclear weapons, and the situation in a state like Bihar, where the private militias of landlords compete with Maoist guerrillas, and the latter compete with each other. Inter-community violence is not easily suppressed by small, high-tech regular forces.

International conflict over resources may also involve attempts to redistribute wealth at a global level. This *could* be presented in terms of the long-cycle theory of international conflict, in which war is related to historical-structural cycles of world politics and economics. Indeed, from the perspective of such analysis, it is misleading to feel confident that transitional warfare amongst the major powers has come to an end.[7] The globalization of economic systems enjoys only limited popular support. Furthermore, it will remain possible to rally domestic support by focusing on foes.

The importance of the issues at stake induces scepticism about claims of the end of war. In addition, such suggestions suffer from a misleading tendency to extrapolate a condition that is seen as characteristic of the great powers, and to argue or imply that other societies can be made to accept their lead. The flavour of such an approach can be seen in the closing passage of a published lecture on the history of warfare by John Keegan:

. . . war *is* now avoidable; war is no longer *necessary*. The poor may fight, but the rich rule. It is with their weapons that the mad ideologies of peasant countries tread the path of blood . . . at the threshold of a new era in history, can we but seize the opportunity, on the threshold of a genuinely new world order. We can stop now if we only choose, by a simple economic decision of the governments of the rich states not to make more arms than they need for their own purposes, and not to supply any surplus that remains to the poor, the have-nots . . . The time has come [to end war.][8]

Irrespective of the ability to secure good relations between the major powers so that they will not attempt to exploit opportunities elsewhere in the world, such hopes rest on a naïve assessment of the relations between the 'rich' and the 'poor'. Descending to specifics, it is far from clear how best to classify oil-rich states, such as Iraq or Libya, that have a revisionist agenda. It is also unclear that limiting the arms trade will prevent the 'poor' from gaining lethal weaponry. The notion that conflict is an 'add-on' that is not integral to human history can be questioned.

In addition, there is no global political agency capable of giving military backing to any contested attempt to control or allocate resources, whether military or other categories of resources. After 1945, it had been hoped that the United Nations would dispose of standing forces. The Military Staff Committee that advised the Security Council sought to agree the allocation of units from the five permanent members of the Security Council, but they were unable to agree and, in 1948, abandoned the task.[9] As a consequence, the UN resorted to ad hoc forces for its operations, while, more generally, aspirations towards global co-operation fell victim to the role of states, the reality of the Cold War, and their own impractibility. These aspirations were important to the terminology of UN operations, encouraging an emphasis on 'peacemaking' not 'war'. This suggested an ability to control the commitment, and a limit of effort that did not disrupt peace or peacetime expectations of the relations between individuals and the state. Such terminology and assumptions could be misleading.

General considerations can be focused on particular states and relationships, for example that between China and the USA. China's prosperity has helped make it more assertive, as has its sensitivity to criticism of its domestic situation. Furthermore, Russian weakness has both made China less vulnerable and decreased its interest in good relations with the USA. Russian financial problems have also encouraged the sale of advanced weaponry to China. This has included missiles, submarines and aircraft. In 1999, there were reports that Russia had agreed to sell two Typhoon-class nuclear-powered submarines. Greater prosperity will help increase China's dependence on imports, and thus its sensitivity to the distribution of resources. It is unclear whether China will learn how to operate, balance and advance its interests within the international community, as powers seeking a long-term and stable position must, or whether it will strive for advantage in a fashion that elicits opposing actions, and possibly war.

The most likely opponent will be the USA, not only because of its specific interests in the West Pacific, and, to a lesser extent, South East Asia, but also due to its more general concern to preserve the international order that it has created. In the case of China, there is also a tradition of American hostility, alongside that of a search for better relations, and this tradition, and its location in terms of the political cultures of the two states, provides both sides with a ready vocabulary for dispute.[10] The 'comprehensive engagement' that America under President Clinton sought with China was designed to limit Chinese revisionism of the international order.

A conflict would test the military effectiveness not only of the USA and China, but also of their allies. Indeed, the nature of a possible great power clash has encouraged American criticism of allied armed forces on the basis of their inadequacy for any such conflict.[11] This theme has been taken up in some allied countries. From both sources, there has been widespread criticism of the nature of European forces, and specifically of their inability to keep pace with American innovation in doctrine, weaponry, organization and expenditure. For example, the lead editorial in *The Times* (London) on 5 August 1999 complained:

277

The European Union spends 60 per cent as much as the US on defence, but packs only a fraction of America's military punch. Far too much goes on conscripts unfit for high-intensity battlefields, expensive facilities, inefficiently organised support costs and unnecessarily expensive weapons procurement; and far too little on advanced military information systems, weapons research and high-quality career-based units.

The assumption that military means should take precedence over political objectives was clear. In fact, expensive weapons procurement, facilities and support costs arose from NATO's character as an alliance in which each state wished to preserve military functions. Similarly, the political reasons for conscription in some states are commonly neglected, as are the political drives behind the run-down in the size of the military in the 1990s.

Yet it is worth bearing in mind that it is by no means clear that the global projection of power that the USA seeks is an objective shared by its allies (or indeed opponents). Allies will be, indeed have been, berated for this, but much of the criticism rests on a failure to appreciate the perspective of particular countries and their individual geopolitical and military traditions and specific political problems. Advocates of the RMA ignore this diversity in favour of a presentation of the world as, in effect, an isotropic surface. It is the suggestion of this book that such an assessment is both politically and militarily naïve. Furthermore, this book seeks to show how this naïvety has a past, not only in terms of past deficiencies in technologically advanced systems, but also in analytical narratives, both present-minded and historical, predicated on technological superiority.

A careful probing of past 'military revolutions' highlights this point, as it indicates how new technologies did not necessarily transform the geopolitical and military traditions and strategic assumptions of particular states, and how the weaknesses of technology remained more important than technological weaknesses. For example, the new land-warfare technology of tank attacks with air support developed in the inter-war period was employed by the Germans to defeat France in 1940, in a fashion that was not dramatically different to the offensive of 1870 and that attempted in 1914. In each case, the use of technology was framed by strategy, not vice versa.[12]

Furthermore, the tank was most effective in co-operation with existing technologies and other arms, rather than on its own. This is a point of direct relevance for the RMA and for future military 'revolutions'. Thus, in the Second World War, tanks were found to be most effective in co-operation with mechanized infantry and mobile artillery. Armoured divisions that were balanced between the arms were effective, rather as the Napoleonic division and corps had been. In addition, the impact of new technologies was limited not only by their diffusion but also by the development of technological, tactical and strategic countermeasures.

Atomic weaponry can also be seen as framed by existing assumptions, although its (fortunately) limited use ensures that it is difficult to write with authority on this point. In one respect, atomic weaponry was treated as an extension of already powerful military organizations and doctrines. First, it was used in 1945 and thereafter envisaged as a form of strategic bombing, although in 1945, General Marshall also considered using atom bombs in tactical support of a landing on Kyushu.[13] Second, the tactical nuclear weapons developed in the 1950s were treated as a form of field artillery. In short, organizations and established strategy moulded the use and planned application of technology, although there was also a major shift in resources from the army towards, first, the air force, and then to submarines and land-based inter-continental missiles capable of delivering atomic warheads.

Against this comes the view that atomic weaponry was so potent that it led to the end of war: in short, that the aggregate capability of the system did lead to a total change in the nature of war. This is an argument that has to be handled with care. The absence of great-power war in 1946–99 was not unprecedented: there was no such war in 1816–53 or 1857–1913 (assuming that Austria, France and Prussia are not regarded as great powers comparable to Britain or Russia; if they are, the dates change to 1816–53 and 1872–1913).

As already suggested in Chapter Eight, technology was subordinated to other considerations, such as imperialism, in the nineteenth century. Second, in 1946–99 the great powers competed through surrogates in the Third World. Thus, *if* nuclear weapons prevented a certain type of warfare, they did not stop conflict. More generally, the implications of nuclear weaponry for the RMA and for the future of war are unclear.[14]

Atomic weaponry also posed, and will continue to pose, the problem of how best to assess capability. This is always a serious problem for military analysts, but is far more so for weaponry that has not been put to the test of conflict. Thankfully, atomic weapons have only been used twice in war. There has been no conflict between nuclear powers, and no employment of the greater, improved post-1945 atomic weaponry. This problem is more generally true of all developments in modern 'advanced' warfare. It is not simply of abstract interest, especially given the destabilizing role of a sense of threat in encouraging the build-up of military force, and even pre-emptive action. For example, in the early 1960s, anxieties about the nuclear balance encouraged Kennedy to aim for a strategic superiority and Khrushchev to decide to deploy missiles in Cuba, thus bringing the world close to nuclear war.[15] Furthermore, Kennedy's increase in defence spending led to a Soviet response that threatened the American position in the 1970s.

It is interesting to note the parallel between atomic and chemical warfare. The failure in the Second World War to use gas and thus fulfil fears raised after the First World War does not mean that there is no danger of chemical warfare today and in the future, although the problem of accurate delivery systems remains. The Sixth Annual Report of the British Chemical Warfare Research Department observed in 1926:

It is an illuminating commentary on the real attitude of the different countries towards conventions and agreements, to observe that practically every one of the countries (some 40 in number) which concurred in the adoption of the chemical warfare prohibition at the League of Nations Conference in June 1925 is actively studying and developing chemical warfare, it is true, in some cases only from the protective point of view, but in most cases both offensively and defensively'.[16]

A Chemical Weapons Convention signed in 1993 that banned the possession of such weapons is, in practice, of only limited value. Similarly, the Non-Proliferation Treaty of 1968 did not fulfil its objectives of limiting nuclear warfare capability. Iraq and North Korea were among its signatories, but both took steps to circumvent its provisions.

It is possible that one of the most important developments in forthcoming decades will not be the use of novel military technologies, but rather their employment by powers that have not hitherto possessed them. Thus, for example, second- or third-rank powers and non-state movements might be tempted to use nuclear, chemical and bacteriological weapons of mass destruction, not least because they are not inhibited by the destructiveness of such weaponry, which is also relatively inexpensive. The nature of international asymmetrical warfare may, therefore, become much more threatening to the more 'advanced' power, and its home base may be subject to attack. Furthermore, it is possible that the RMA will not overcome this new vulnerability, although, in turn, this vulnerability helps to drive investment in new technology, just as it also encourages the maintenance of the American strategic nuclear arsenal.[17]

Again, the possibility of conflict with China offers a way to probe the future dimension of these issues. It is by no means clear that such conflict would necessarily have the clear origins – the Pearl Harbor scenario – that would permit the presentation of China as aggressor, and thus justify a call on civilian and allied resources. Even if it does, if advanced weaponry is used in all-out conflict, it is not clear whether it could fulfil objectives or survive a rapid depletion rate. The timetable of conflict by the two sides might be very different, and such that the short-term high-intensity use of advanced technology by the Americans achieves devastating results in 'deep battle', but without destroying the Chinese military system or the political determination to refuse American terms.

A similar challenge will affect the USA in the event of any clash with major Islamic powers; the Gulf War is not a clear guide because the USA was aligned with Islamic states and there was no attempt to overrun Iraq. The military challenge posed by Islamic assertiveness, whether fundamentalist or not, has been underrated, because there is no single major Islamic state comparable in power to India. Furthermore, there are powerful rifts within Islam, not least between Shias and Sunnis. This helped strengthen antagonism between Iran and Iraq in the 1980s, and Iran and Afghanistan in the 1990s.

However, it is anticipated that by 2030 Muslims will make up about 30 per cent of the world's population, and it has been claimed, most prominently by Samuel Huntington, that religious and cultural clashes and fault lines are the most probable cause of conflict after the end of the Cold War. The last twenty years have shown that Islam has become more resurgent and more of an issue across much of the world, not least in Central and South Asia, sub-Saharan Africa and the Balkans. That does not imply that Islam is a united force, indeed far from it, but nevertheless, there are common themes of assertiveness. Much of this had involved a rejection of the civilization and values associated with the West, as well as its political and economic views. Thus, in the event of conflict, there is an ideological clash that will make moderation in conflict and compromise in peacemaking difficult. The experience of the Arab–Israeli wars, where the Western-equipped and organized Israeli military found it difficult to triumph in the conflict, as opposed to emerging victorious from individual wars, particularly that in 1967, focuses this problem, although a number of factors, including the emotive character of religious sites and the propinquity of the combatants and their peoples, made the issues particularly pressing.

More generally, the contrast between American and Islamic attitudes underscores the difficulty of using the United Nations. The divergence in views towards the objectives of government, the nature and value of human rights, and the use of force among the member states is a challenge both to the future operation of UN forces and to the ability to agree goals. For a long time, this problem was in part masked by Cold War divisions, but it is now clear that the problem is much more profound and also more complex in character. An unwillingness to accept American doctrine, strategy and tactics is likely to affect many UN operations in the future, irrespective of whether the American identification of foe and interpretation of peacekeeping is accepted. Indeed, in Vietnam, American reliance on firepower led the Australians, who had a different notion of jungle warfare, to seek (and obtain) their own tactical area of responsibility.

The elusiveness of lasting success and the contrast between it and what is termed total victory brings into focus a central problem. The RMA is essentially another guide to military success, but like the German Schlieffen Plan of 1914 and the *Blitzkrieg* employed by the Germans in 1939–41 and the Israelis in 1967, it fails to explain how those who have been defeated will be persuaded to negotiate. In short, victory is useful, but no more, unless the defeated are prepared to accept the verdict. High-intensity warfare – rapidly resolved clashes of conventional weaponry – will be succeeded by more intractable problems. Symmetrical and asymmetrical conflict will also not necessarily be distinct. Indeed, their simultaneity in individual wars may exacerbate the difficulty of both.

Yet it is possible that such pragmatic considerations are beside the point. To treat the RMA as a technical discussion of military capability is to ignore the psychological factors that encourage belief in it. First, as already suggested in the previous chapter, it meets needs for a belief in superiority. Whereas, for many other cultures, and in the past, for Western culture, this has generally been expressed in terms of confidence in morale, such as Japanese *bushido*,[18] over the last fifty years it is technology that has overwhelmingly taken this role in the West. It is, apart from anything else, more appropriate for the deracinated societies of the modern West than an emphasis on moral characteristics, which frequently had racist connotations.

Second, the expression of a quest for superiority in technological terms meets the imaginative demands of modern industrial society. This was especially important in the case of air power, which powerfully gripped the inter-war imagination of the West,[19] and led to much anxiety about the likely impact of the bomber in wartime.[20] In practice, the inter-war use of air power had already suggested that it might be less effective than its protagonists claimed,[21] and, as far as Britain was concerned, the Air Ministry was more moderate in its claims than independent air enthusiasts.[22] Nevertheless, the notion of a paradigm leap forward in military capability, and of a resulting situation of clear-cut advantage over others remained attractive, and was translated to post-1945 technologies.

Thus, the RMA can be seen as serving powerful psychological needs. It is crucial to Americans that this is an American military revolution, not least because it permits a ranking in which America is foremost and all other powers, whether opponents or allies, are weak and deficient. The RMA can serve the doctrine, politics and military strategy both of American isolationists and of believers in American-led collective security. The RMA meets the American need to believe in the possibility of High Intensity Conflict and of total victory, and appears to counter the threats posed by the spread both of earlier technologies, such as long-range missiles and atomic weapons, and of new ones, such as bacteriological warfare.

Indeed, all these threats are present in the event of a confrontation or clash between the USA and its allies, and North African and Middle Eastern states (including Iran), let alone China. Their rapid development of multi-stage rockets and of a range of non-conventional warheads poses acute problems that are exacerbated by the relative volatility of several of these states. In comparison to the leadership of Libya or Iraq in the 1980s and 1990s, that of the Soviet Union in the 1960s, 1970s and early 1980s, in hindsight seems *relatively* predictable and cautious in its policies. These military threats are the counterpart of possible political and economic challenges for the reconfiguration or redistribution of global power and resources.

The notion of an RMA can be seen differently if it is presented as a doctrine designed to meet political goals by countering political challenges, and thus to shape or encourage technological developments and tactical suppositions, and contrasted with the idea that technological constraints should shape doctrine, and thus inhibit policy. It is possible to take very varied positions on the merits of these arguments, as can be seen by probing historical examples, such as that of the use of strategic air power in the Second World War.[23]

The notion of a leap forward was also, and continues to be, the stuff of futuristic writing about war. This is especially obvious in American works. For example, David Alexander's *Tomorrow's Soldier: The Warriors, Weapons, and Tactics that Will Win America's Wars in the Twenty-first Century* (New York, 1999), a populist account of such technology, includes 'Scenario: Battlezone 2010'. This begins: 'Events now gather increased momentum . . . At the port city of Shabaz, F-1/7B *Nighthawks* destroy major communication nodes with precision laser-guided bomb strikes', and so on. The focus throughout is on fighting, not its wider context in conflict. This approach draws much of its popularity from the presentation of science fiction in terms of destructive conflicts with aliens which involve high-tech fights to the finish. Low-tech conflict, and postwar pacification and policing, do not play a role in much of this, although there are some science fiction series, for example *Star Trek*, that do attempt to probe antagonistic relations short of all-out war.

Much military doctrine is similar to futuristic writing about war. Advocates of the RMA have progressed to talking about 'space control' and the 'empty battlefield' of the future, where wars will be waged for 'information dominance' – in other words, control of satellites, telecommunications and computer networks. The US military in the 1990s referred to 'information grids' that must be safeguarded in wartime, and hostile ones that must be destroyed, as those of Iraq (1991) and Serbia (1999) were. The US planned for four interlocking grids: a communications grid, a sensor grid, an engagement grid, and a defence suppression and protection grid.

Future tactics are increasingly discussed in 'joint' terms, as in the American *Joint Vision 2010* plan. Indeed, there has been a proliferation of 'joint' organizations – in Britain, for example, the Joint Rapid Deployment Force, the Permanent Joint Headquarters and the Joint Services Command and Staff College, or in the USA the creation in 1992 of an Expeditionary Warfare Division in the office of the Chief of Naval Operations, and more generally, the Goldwater-Nichols Department of Defense Reorganisation Act of 1986, which strengthened the position of the Chairman of the Joint Chiefs of Staff and established a joint acquisition system. Joint institutions provide powerful advocates for new doctrine and planning. This process has led to discussion of the obsolescence of traditional distinctions between the services, and indeed of the notion of separate services. This may well be true of battle between conventional forces, but there is a world of difference between the task of counter-insurgency operations on urban streets and long-range missile attack on hostile states.

American economic growth in the 1990s made it easier to think of investing in new systems. In planning for future weaponry and warfare, there is a tendency towards cheaper, unmanned platforms, such as UAVs (unmanned aerial vehicles, the same as RPVs – remotely piloted vehicles) to replace reconnaissance and attack aircraft, 'jumbo cruise missiles' to replace Tomahawks, cruise missiles to replace tanks, and 'arsenal ships' to replace the carrier battle group. An emphasis on precision is designed to permit the definition and differentiation of civilian from military targets. Area firebombing is out. There is an emphasis on avoiding risk for troops, and thus on unmanned weapon platforms. As in the past, particular weapons systems and strategies are advocated and discussed in terms of doctrines that reflect the composition and culture of particular military institutions.[24] The quest for more modern hardware is very expensive, and not always well informed by issues of practicality and likely political role.

As generally discussed, the RMA is a means to total victory, providing the opportunity for a universal warfighting doctrine that offers very little role for the Low Intensity Conflict that was the combat norm during the Cold War and has become even more so subsequently. In short, there is a danger that the RMA serves as an apparent substitute for a political willingness to commit troops, and also as a cover for the failure, first, to develop effective counter-insurgency doctrine and practice,[25] second, to conceive of a strategy for successful long-term expeditionary operations, and third, to work within the difficult context of alliance policy-making and strategic control. It is far from clear that modern Western military systems (let alone their politicians and public opinions) are ready for the long-term policing of recalcitrant populations. It is at this point that the ability to create tactical and strategic countermeasures – the petrol bomb, guerrilla warfare or civil disobedience – to cope with technological deficiencies is most apparent. Nor is it necessarily the case that the technologically more advanced society is better able to respond to and learn from its opponents' fighting methods than vice versa.

Nevertheless, both military and political pressures in the modern world encourage concern and confrontation with, even intervention in, other states, including those at a considerable distance. The range of readily available modern weaponry is such that it is difficult to ignore hostile developments elsewhere, specifically the production of weapons of mass destruction. The perception of social threats will also lead to military interest in distant regions, as with the American concern to prevent the production of narcotic drugs in Latin America. This attitude may be extended with environmental concerns. Long-range, and even universalist, conceptions of American interests will interact with the global pretensions of the United Nations and the sense that the settlement of international problems is a responsibility for neighbouring and other powers. The resulting premium on order, of a certain type, will encourage concern and insecurity, and will be difficult to enforce.[26] Thus, one impact of globalization – namely, the wider implications of local struggles – is potentially destabilizing.

There has been talk of unconventional methods to defend Western values, including 'psychological operations', 'mental intrusion' and non-lethal technologies, but it is far from clear that these will be effective. Non-lethal weaponry includes anti-personnel weapons across a wide range, from psychological, visual and sound to physical and chemical, and anti-material ones, such as computer viruses, adhesives and metal embrittlements. Some of these weapons are relatively well known, such as rubber bullets and tear gas, while others are more advanced, such as chemical incapacitants. In addition, some of the latter have been used, for example the 'foam' and 'toffee' guns that the US Marine Corps took to Somalia, and the carbon filaments fired by the Americans to short-circuit Iraqi electrical plants. Others have not been employed in conflict.[27]

The problem of eliciting consent facing America and the UN can also be applied to more modest state-to-state clashes and intra-state warfare in the future. Both face the problem of enforcing a verdict and, more generally, of the wasting quality of international order. This quality requires consent for its maintenance, rather than force. Force in the absence of consent is of limited value. That is not intended as an anti-war remark, because the nature of communities and international systems is such that consent alone cannot be relied upon. However, to move to the opposite position and assume that force can operate without an attempt to build up consent is of limited value. Indeed, the experience of the last 150 years suggests that those brought low by force have an ability to reverse the verdict, either peacefully or by violence. As a consequence, the most sensible policy – seen, for example, in the treatment of West Germany, Japan, Italy and Austria after the Second World War – is the attempt to rebuild a civil society to whom authority can be entrusted; yet such a policy requires a total victory and entails a massive postwar commitment by the victor. To achieve either this end or the more credible goal of negotiations based on a more limited success, it is best to wage a war in which the opposing society is not demonized. Part of the problem in the 1990s, in the case of Western confrontations with both Iraq and Serbia, is that an adequate exit strategy from the war was not developed.

Liberal internationalism encouraged a globalization of commitment from the 1940s that was further enhanced by the global character of Western capitalism. However, it remains unclear whether this internationalism was based on a sensible assessment of means and goals. On one level, such an assessment can indeed be found in the restraint displayed by the cartography of Western concern. There was no war to drive China from Tibet, no intervention to try to resolve conflicts in the Caucasus or Central

Africa. The reality of a new world order was thus far more restrained than the language. Liberal internationalism also encountered the problem that the major Western power, the USA, had an ambivalent relationship with the constraints of collective security, especially with the United Nations, and not least with the notion of UN direction of operations involving American forces. Furthermore, military procurement policies continued to suggest a limited range for large-scale deployment of ground forces by every power other than the USA. This was shown, for example, in major differences in the amount of strategic lift made available by outsize cargo-carrying capability, such as the American Starlifter aircraft.

On the other hand, the wars that were waged left the Western powers with unresolved problems and large military commitments. Both potentially limited their ability to respond to fresh problems, as seen with American anxiety (which will no doubt persist) that a conflict in the Balkans or the Middle East might affect the response to a crisis in East Asia, specifically over South Korea or Taiwan. In 1999, the USA sent the sole aircraft carrier permanently assigned to the Pacific to the Mediterranean during the Kosovo Crisis. Force reductions after the end of the Cold War exacerbated this problem and made it more likely that a major commitment would entail coalition resources. President George Bush's intention that the USA would be able to launch two DESERT STORM-scale operations concurrently has had to be abandoned. Irrespective of this, the Americans funded DESERT STORM in part from allied financial contributions, particularly those of Saudi Arabia, Japan, Germany and Kuwait. Yet there are tensions between the external constraints that alliance policy-making entails and the nature of political culture in the USA, which tends to be hostile to compromise with foreign powers. American politicians and public opinion instinctively think in unilateral, not multilateral, terms.

Similarly, Russian commitments in particular areas, such as Chechnya or Dagestan, may limit their potential to act elsewhere, for example in Central Asia. There is also the danger that the range of possible commitments will lead to an overly strong (or rapid) response in one particular crisis in the hope that this will affect other

289

possible zones of tension. It is generally harder to control the consequences of this prophlyatic use of force than might at first appear. This can be seen as an aspect of the Russian response to crises over Chechnya, although the nature of the response also reflected the Russian preference for firepower and, more generally, the dominance in doctrine and practice of preparations for the Cold War rather than low-intensity and counter-insurgency warfare. As with the USA, this is an aspect of the continued impact of Cold War military culture, although there are other factors. For example, the emphasis on force and response doctrine and tactics in American and Russian policing is not that of neighbourhood policing, with its stress on co-operation with local communities.

The current state of the Russian armed forces acts as a brutal reminder of the wider contexts of military power. It was not failure in Afghanistan that was responsible for their crisis in the 1990s, but rather the political disintegration of the Soviet Union and the governmental and financial weaknesses of Russia. The state could no longer support its forces, and they became weak and divided, and thus far less able to maintain their effectiveness or to fulfil military functions.

The abandonment of Communism brings into focus the problem of how best to classify Russia as a military power, not least in terms of the West–Rest continuum. Whatever the classification, it would be mistaken to fall into the trap of focusing on conflict between the West and the Rest. As a consideration of past and present conflict indicates, it is necessary to note the importance of West v. West and Rest v. Rest. To take China, it may well be the case that, as in 1500–1830, the principal military challenges come not from other Western powers, but rather from within China and along its non-Russian land borders. For example, separatism in Xinjiang may interact with intervention in the Central Asian republics. Alternatively, China may be drawn into conflict between Vietnam and its neighbours, as in 1979, or between India and Pakistan, or into Burma. Furthermore, these commitments might not involve any direct clash with the West.

The Chinese do not have a tradition of decolonization to which they can turn to justify withdrawal from Tibet or Xinjiang, and instead reject a definition of China restricted to the Han Chinese legacy. This creates serious problems in terms of an ability to cut short difficult military and political commitments. In addition, unlike Britain from the late 1940s, China does not have a powerful ally able to reassure it about its defensive needs (not least by locating forces on its soil), and also to encourage decolonization. It is unclear how far the relevant example of Chinese success in border conflict will be the Korean War, the Sino–Indian war of 1962 or that with Vietnam in 1979. The last showed China's willingness to take losses – 46,000 casualties in seventeen days out of the 250,000 troops deployed.[28]

This stress on border confrontations with non-Western powers is not restricted to China. Iranian commitments in the future may be in, or opposed to, the Caucasus, Central Asia, Afghanistan, Pakistan or Iraq, and not at the expense of oil-rich states along the southern shore of the Persian Gulf that are linked to the West. Similarly, Iraq could be most concerned about Iran or the Kurds. Each has involved Iraq in conflicts far more protracted than the Gulf War. Syria has fought Israel in conventional state-to-state warfare, but also intervened in complex struggles in Lebanon, while in 1982 a full-scale Sunni insurrection in Hama was crushed. This analysis is not only appropriate for Asia, but also for Africa. If, in addition, Cold War links are relaxed, then it will be increasingly possible to see such confrontations and clashes as unrelated to any West v. Rest alignment.

As suggested in earlier chapters, the West v. Rest approach can be extended to consider internal challenges. The latter also offer the best way in which to assess West v. West in the future. Formal conflict between states is unlikely, not least because of economic links and American military, political and economic hegemony. However, conflict within states is increasingly likely, as governments find it harder to persuade minorities to renounce violent opposition. This involves a counterpart to the emphasis on technology in international warfare. In theory, modern states are far better able to control and suppress discontent. They have the capacity to create a surveillance society in which the government possesses considerable information about every individual. Furthermore, the nature of the modern salaried workforce and (through social security) non-workforce is such that most people cannot afford to break from this surveillance society. Aside from information, modern Western governments also have massive resources. In addition, their internal control forces, whether military or police, have communications and command and control facilities that are far greater than those enjoyed even twenty years ago.

Yet internal policing has also been greatly affected by the RAM. This is likely to continue to be the case, and will probably be accentuated by the breakdown of respect for government, and a process of social atomization that potentially leaves common enthusiasms as the major way in which people are motivated. These enthusiasts will, in some cases, be unwilling to accept the disciplines of democracy: subordination to majority opinions, and mutual tolerance against the background of the rule of law. It is unclear that states, whatever their capacity for surveillance, will be able to suppress the resulting violence, in part due to the difficulty of the task, and in part a response to the constraints affecting their response. Peacekeeping as a military task will interact with what has been seen as the breakdown, or at least reconceptualization, of the state, a concept frequently expressed by referring to the post-Westphalian state/system. The inability of the British government to disarm terrorist groups in Northern Ireland or of the Spanish government to bring peace to the Basque region does not augur well.[29]

Looking to the future, it is difficult to see any abandonment in developed economies of the notion of a specialized military.[30] The RMA, in this respect, is simply another stage in the move away from the notion and practice of the citizenry under arms. Indeed, it is part of the search for continued military potency by a society that can no longer countenance the mass mobilization and ideological and social militarism that characterized its conflict with totalitarian regimes in the Second World War and the first two decades of the Cold War. To defeat Germany and Japan and to forestall Communism, it proved necessary to match the impressment and ideological coherence (at least over foreign policy) shown by these armed nations. Yet both were inappropriate for American social politics from the 1960s. As a consequence, the notion of the armed nation had to be reconceptualized, back towards professionalism, with the citizenry paying the bill through taxation. This shift matched American political needs and military doctrine in the post-Vietnam Cold War. It also ensured that a diminishing percentage of the population, the electorate, the taxpayers and the politicians had seen military service and thus (on the whole) been habituated to military assumptions.

A true military revolution would be a return to the citizenry under arms. It is definitely not necessary in terms of modern military technology. For example, Iran, a state that has the ethos necessary for mass mobilization, as it showed in its 1980–8 war with Iraq, is currently developing or purchasing and integrating specialized high-tech capability, such as missiles and submarines. Furthermore, the *likely* speed of a future major war is such that there would not be the time to move from specialized to mass forces. Nor would there be sufficient advanced weaponry. For example, a future conflict between Iran and Iraq, when between them they possessed atomic, chemical and bacteriological weaponry, would probably be over more rapidly if they employed this weaponry than their 1980–8 war, although the subsequent 'pacification' stage of the power defeated in this clash might be far more lengthy. In addition, the example of the Second World War, which was far lengthier and involved far more troops than the advocates of mechanized[31] and air warfare in the 1930s had anticipated, suggests that a degree of caution about the future manpower implications of major conflict may be appropriate.

More generally, both World Wars suggested that the effectiveness of new technology-based strategies was less than might have been anticipated even prior to the attritional stages of the conflict. Thus, the railways that were so useful during mobilization in the First World War proved of limited value as forces advanced into hostile territory. In the Second World War, the potential petrol engine-based weaponry and logistics were less fully graspable by the Germans than talk of the *Blitzkrieg* might suggest. The availability and movement of petrol posed problems for all powers.[32] It is probable that a similar situation will pertain in the future.

Mobility is also crucial in dealing with domestic problems – indeed, more so. The decline in the size of armed forces is less of a problem when engaging similar units than it is when attempting to control a large civilian population. Mobility is an important force-multiplier in the latter task. Thus, for example, good road links were important to empires such as the Aztec, Chinese and Roman. These links did not, however, provide a clear advantage over opponents, who could also march or ride. Such a gap developed with the railway, although only in so far as the network extended. Road links and air power were more extensive. A sense of new potential can be seen in British staff reports of the 1920s. For example, the *Military Report on the Sudan* prepared by the General Staff in 1927 commented on the value of the armoured cars of the three motor machine-gun batteries in the Sudan Defence Force: 'so that small local disturbances may be quelled before they can spread'.[33] Two years earlier, the British had begun to use both night bombing and delayed-action bombs to limit tribal hostility in Waziristan.[34] The Internal Security Instructions for India issued by the General Staff Branch in 1937 emphasized the combination of firepower and mobility:

Automatic weapons have rendered easier the defence of keeps, posts, important buildings and bridges, etc., from attacks by insurgents. The number of troops employed on defensive duties can be reduced accordingly. The measures taken by troops should be active rather than passive. Patrols and mobile columns are more efficacious than fixed piquets and defensive measures for quelling disturbances. . . . The number of troops employed on purely defensive measures should be reduced to a minimum, and strong mobile reserves should be formed. The fullest possible use should be made of the mobility conferred by motor and air transport to reduce the number and size of detachment.[35]

Yet the effectiveness of such tactics in the modern, increasingly urbanized world is limited. Instead, urban areas are difficult to control, and mobility is affected by the nature of the built environment. This is likely to become a more limiting factor as cities sprawl and recognizable centres of control or influence (such as television stations) become more numerous or can be readily duplicated. The contrast between conventional operations and those in divided urban environments was captured by British troops flown from Italy to Athens in December 1944 in order to contain ELAS, the left-wing Greek People's Army of Liberation. A signaller recalled:

> The enemy were just the same as any other Greeks as far as we knew, they didn't have any uniform as such. . . . It was a situation that was quite completely different to the way we had been used to fighting. . . . As an average infantryman, one of the first questions that you ask . . . is 'Which way is the front?' So that you know if the worst comes to the worst which way you can go to get out of the bloody place. In this sort of situation, which is a typical urban 'battlefront' it's all around you and one feels somewhat unhappy about that.[36]

At present, it is probable that there will be numerous wars in the future, both short- and long-term, and that in them the language of total commitment would be employed to explain, justify and encourage economic direction and the retrenchment of consumption, rather than the mass mobilization of potential combatants. This may, however, be a mistaken assessment because confrontations (as opposed to war) could be very lengthy, while economic and political disruption might encourage the recruitment into low-level units of the potentially disaffected. Furthermore, the use of the military for policing units requires a higher density on the ground, and thus more men, as well as less skill, advanced weaponry and expenditure, than their employment in conflict with regular forces.

The importance of maintaining 'order' as a military task may well be underrated because the major Western power, the USA, has not faced a convincing separatist or terrorist movement. Furthermore, violent or potentially violent groups in the second half of the

295

twentieth century, such as the Black Panthers, the Klu Klux Klan and the Militias, were all contained.

It is unclear how far it is appropriate to be optimistic about the future. On the one hand is the powerful incorporating character of American society and the American myth, the success of both, and the atomistic and poorly organized nature of their opponents. The success in integrating the American military, lessening institutional racism, and improving conditions for African-Americans is notable.[37] On the other, as a recent historian of the USA has suggested, the Oklahoma bombing in 1995 'represented – albeit in a dark form – some American traditions that a government neglects at its peril: radical individualism, suspicion of government, apocalypticism and a willingness to strike back at perceived tyranny'.[38] How far this should be presented as a form of war captures the difficulty of assessing the term. The degree of armaments on both sides – as seen, for example, in the siege and storming of the Branch Davidian compound at Waco in Texas in 1993 – was such that paramilitary action rather than policing is the concept that comes to mind. This is not too different to the situation in many other parts of the world.

Given the ability of American governments to deploy paramilitary units rather than the regular armed forces, and given the depoliticized nature of the latter, it is probable that the USA will be able to continue to regard internal violence as separate from war and, crucially, not part of the function of its military. This again captures American military exceptionalism and helps to explain why the Americans underrate the infantryman, the most flexible tool of war and the prime point of contact between responses to external confrontation and to internal challenges. It also conditions the hostile American response to the use of the military to maintain control and 'order' in other states, whether by America's allies, by its opponents, or by other powers.

Another aspect of variety is captured by the issue of the flexibility of particular military forces, specifically their responsiveness to the tasks they face and to developments in warfare. There is an institutional dimension to this question, not least the degree of decentralization in command structures and ethos, but it can also be seen as at once an aspect and a product of the characteristics of different societies. The USA is far more flexible than autocratic societies, both in considering past and present and in searching for advantage in the future, although this flexibility can lead to a serious lack of consistency in foreign policy.

A society like the USA, which is open to talent and without a caste-like social structure, a rigid ideology or an autocratic government, should be better able to respond to the need for initiative in military doctrine, strategy and tactics. In both the Second World War and the Cold War, the Americans were more successful than their opponents in encouraging and organizing a systemic productivity that provided a massive build-up of an effective military[39] without jeopardizing the domestic economy. It is unclear, however, how far this will in the future provide a military edge that can be employed successfully to solve problems, and it is necessary to remain aware of deficiencies in military systems in order both to improve their effectiveness and to understand their capability.[10] The latter has to focus on the contrast between successful campaigning and the achievement of policy aims.

To conclude on a note of future variety and likely difficulties is to suggest that attempts to assess military forces and campaigns, capability and effectiveness on a global ranking will be flawed in the future as they have been in the past. Instead, an understanding of the limits of technology, of counterfactual arguments and of the role of struggles within states are all important in attempting to assess the future military history of the world.

NOTES

1. J. Mueller, *Retreat from Doomsday: The Obsolescence of Major War* (New York, 1989); C. Kaysen, 'Is War Obsolete? A Review Essay', *International Security*, 14 (1990), pp. 42–69; R.L. O'Connell, *Ride of the Second Horseman: The Birth and Death of War* (New York, 1997); M. Mandelbaum, 'Is Major War Obsolete?', *Survival*, 40 (Winter 1998–9), pp. 20–38.
2. E. Karsh, 'Cold War, Post-Cold War: Does it make a difference for the Middle East?', *Review of International Studies*, 23 (1997), pp. 271–91.

3. M. Kidron and R. Segal, *The State of the World Atlas* (1995).
4. P. Kennedy, *Preparing for the Twenty-first Century* (New York, 1993); R. Kaplan, 'The Coming Anarchy', *Atlantic Monthly* (February 1994), pp. 44–76; T.F. Homer-Dixon, 'On the Threshold: Environmental Changes as Causes of Acute Conflict', and 'Environmental Scarcities and Violent Conflict: Evidence from Cases', in S.M. Lynn-Jones and S.E. Miller (eds), *Global Dangers* (Cambridge, MA, 1995).
5. J.S. Levy, 'Contending Theories of International Conflict', in C. Crocker and F. Hampson (eds), *Managing Global Chaos* (Washington, DC, 1996), p. 18.
6. L.H. Keeley, *War Before Civilization: The Myth of the Peaceful Savage* (Cambridge, MA, 1996).
7. K.A. Rasler and W.R. Thompson, *The Great Powers and Global Struggle, 1490–1990* (Lexington, KY, 1994), p. 191. For a very different emphasis, see M. van Creveld, *The Rise and Decline of the State* (Cambridge, 1999), pp. 337–54.
8. J. Keegan, *A Brief History of Warfare – Past, Present, Future* (Southampton, 1994), p. 15.
9. R.M. Connaughton, *Peacekeeping and Military Intervention* (Camberley, 1992), pp. 8–9.
10. T.J. Christensen, *Useful Adversaries: Grand Strategy, Domestic Mobilization, and Sino–American Conflict, 1947–1958* (Princeton, NJ, 1997).
11. See, for example, D.C. Gompert, R.L. Kugler and M. C. Libicki, *Mind the Gap: Promoting a Transatlantic Revolution in Military Affairs* (Washington, DC, 1999). This closed with the injunction to the USA to 'extend an offer of partnership in making the RMA a transatlantic endeavour'.
12. J. Corum, *The Roots of Blitzkrieg* (Lawrence, KS, 1992); R.A. Doughty, *The Breaking Point: Sedan and the Fall of France, 1940* (Hamden, CT, 1990), and 'Myth of the *Blitzkrieg*', in L.J. Matthews (ed.), *Challenging the United State Symmetrically and Asymmetrically: Can America be Defeated?* (Carlisle Barracks, PA, 1998), pp. 57–79.
13. Gallicchio, 'After Nagasaki: George Marshall's Plan for Tactical Nuclear Weapons in Japan', *Prologue*, 23 (1991), pp. 396–404.
14. L.W. Martin (ed.), *Strategic Thought in the Nuclear Age* (Baltimore, MD, 1979); H.R. Borowski, *A Hollow Threat: Strategic Air Power and Containment Before Korea* (Westport, CT, 1982); L. Freedman, *The Evolution of Nuclear Strategy* (2nd edn, 1989).
15. M.J. White, *The Cuban Missile Crisis* (Basingstoke, 1995).
16. PRO WO 33/1128, pp. 22–3. See, more generally, R.M. Price, *The Chemical Weapons Taboo* (Ithaca, NY, 1997).
17. O. Thränert, 'Can Nuclear Weapons Deter Biological Warfare?', in D.G. Haglund (ed.), *Pondering NATO's Nuclear Options: Gambits for a Post-Westphalian World* (Kingston, ON, 1999), pp. 81–105.

18. L.A. Humphreys, *The Way of the Heavenly Sword: The Japanese Army in the 1920s* (Stanford, CA, 1995).

19. R. Wohl, *A Passion for Wings: Aviation and the Western Imagination* (New Haven, CT, 1994); P. Fritzsche, 'Machine Dreams: Airmindedness and the Reinvention of Germany', *American Historical Review*, 98 (1993), pp. 685–709.

20. U. Bialer, *The Shadow of the Bomber: The Fear of Air Attack and British Politics, 1932–1939* (1980).

21. D.E. Omissi, *Air Power and Colonial Control: The Royal Air Force 1919–1939* (Manchester, 1999).

22. M. Smith, *British Air Strategy Between the Wars* (Oxford, 1984), pp. 83–4. For the USA, J.P. Tate, *The Army and Its Air Corps: Army Policy Towards Aviation, 1919–1941* (Maxwell Air Force Base, AL, 1998).

23. See, for example, C.C. Crane, *Bombs, Cities and Civilians – American Airpower Strategy in World War Two* (Lawrence, KS, 1993).

24. M. Worden, *Rise of the Fighter Generals: The Problem of Air Force Leadership, 1945–1982* (Maxwell Air Force Base, AL, 1998).

25. D.M. Shafer, *The Failure of US Counter Insurgency Policy* (Leicester, 1988).

26. R.H. Johnson, *Improbable Dangers: U.S. Conceptions of Threat in the Cold War and After* (New York, 1994); W.J. Durch (ed.), *UN Peacekeeping, American Policy and the Uncivil Wars of the 1990s* (Basingstoke, 1996); R.J. Lieber (ed.), *Eagle Adrift: American Foreign Policy at the End of the Century* (New York, 1997).

27. A. Truesdell, *The Ethics of Non-Lethal Weapons* (Camberley, 1996), pp. 14–18.

28. S.J. Hood, *Dragons Entangled: Indochina and the China–Vietnam War* (Armond, NY, 1992).

29. M. van Creveld, *On Future War* (1991); G.M. Lyons and M. Mastanduno (eds), *Beyond Westphalia? State Sovereignty and International Intervention* (Baltimore, MD, 1995).

30. For an emphasis on the continued need for states to be able to wage conventional warfare, see J.A. English, *Marching Through Chaos: The Descent of Armies in Theory and Practice* (Westport, CT, 1996), pp. 193–7.

31. P. Harris, *Men, Ideas and Tanks: British Military Thought and Armoured Forces 1903–1939* (Manchester, 1995).

32. Report of the Standing Committee on the Method of Delivery and Distribution of Petrol to Units in Advance of Railhead, 1936, PRO WO 33/1444.

33. PRO WO 33/2764, p. 257.

34. Summary of Events in North West Frontier Tribal Territory, 1925, PRO WO 33/1135, pp. 12–16.

35. PRO WO 33/1493, p. 3.

36. P. Hart, *The Heat of Battle: The 16th Battalion Durham Light Infantry – The Italian Campaign, 1943–1945* (1999), pp. 201–2.

299

37. J.E. Westheider, *Fighting on Two Fronts: African Americans and the Vietnam War* (New York, 1997); S. Mershon and S. Schlossman, *Foxholes and Color Lines: Desegregating the U.S. Armed Forces* (Baltimore, MD, 1998).
38. P. Jenkins, *A History of the United States* (1997), p. 297.
39. H. Boog (ed.), *The Conduct of the Air War in the Second World War – An International Comparison* (Oxford, 1992).
40. For criticism of a failure to do so, see D.M. Drew, 'U.S. Airpower Theory and the Insurgent Challenge: A Short Journey to Confusion', *JMH*, 62 (1998), pp. 809–32, esp. pp. 824, 829–30. Different countries have changed their airpower 'regimes' – structures and operational dispositions – unevenly. What is 'out of date' to one country and one region retains utility to another country in a different region.

Selected Further Reading

Given the breadth of the subject, it is sensible to concentrate on recent books. Earlier works and journal literature can be approached through the bibliographies and footnotes in these works, although for journals it is worth turning at once to *The Journal of Military History* and *War in History*. Two relatively recent edited works that complement each other are Geoffrey Parker (ed.), *The Cambridge Illustrated History of Warfare* (Cambridge, 1995) and Charles Townshend (ed.), *The Oxford Illustrated History of Modern War* (Oxford, 1997). A less Eurocentric account is offered in J.M. Black, *War and the World 1450–2000* (New Haven, CT, 1998) and (ed.), *War in the Early Modern World 1450–1815* (1999).

A wide-ranging, although controversial, single-volume account is offered by John Keegan, *A History of Warfare* (1993), and from the perspective of a different speciality, J. Diamond, *Guns, Germs, and Steel: The Fates of Human Societies* (New York, 1997). Other fruitful general histories include W.H. McNeill, *The Pursuit of Power: Technology, Armed Force and Society since AD 1000* (Oxford, 1983), and W. Murray, A. Bernstein and W.M. Knox (eds), *The Making of Strategy: Rulers, States and War* (Cambridge, 1994). For Europe, see M. Howard, *War in European History* (Oxford, 1976), and M. van Creveld, *Technology and War from 2000 BC to the Present* (New York, 1989).

Conflict prior to 1500 is not the subject of this book, but among the extensive literature, it is worth looking at A. Ferrill, *The Origins of War* (New York, 1985), V.D. Hanson, *The Western Way of War: Infantry Battle in Classical Greece* (New York, 1989), J. Haldon, *Warfare, State and Society in the Byzantine World 565–1204* (1999), and J. France, *Western Warfare in the Age of the Crusades 1000–1300* (1999).

For the early-modern period, see J.A. Lynn (ed.), *Tools of War: Instruments, Ideas, and Institutions of Warfare, 1445–1871* (Urbana, IL, 1990), G. Parker, *The Military Revolution: Military Innovation and the Rise of the West, 1500–1800* (2nd edn, Cambridge, 1996), D.M. Peers (ed.), *Warfare and Empires: Contact and Conflict Between European and Non-European Military and Maritime Forces and Cultures* (Aldershot, 1997), A. Starkey, *European and Native American Warfare, 1675–1795* (1998), J.M. Black (ed.), *European Warfare 1453–1815* (1999), R. Harding, *Seapower and Naval Warfare, 1650–1830* (1999), and J. Glete, *Warfare at Sea, 1500–1650: Maritime Conflicts and the Transformation of Europe* (2000). A very different perspective is provided by R. Murphey, *Ottoman Warfare, 1500–1700* (1999) and J. Thornton, *Warfare in Atlantic Africa, 1500–1800* (1999).

For the eighteenth century, see J.M. Black, *European Warfare 1660–1815* (1994) and *Warfare in the Eighteenth Century* (1999), and H.M. Ward, *The War for Independence and the Transformation of American Society* (1999). For the development of Western ideas, see A. Gat, *The Origins of Military Thought* (Oxford, 1989) and *The Development of Military Thought: The Nineteenth Century* (Oxford, 1992). For European imperialism in the nineteenth century, see B. Vandervort, *Wars of Imperial Conquest in Africa 1830–1914* (1998). The situation within the West can be approached via G. Wawro, *Warfare and Society in Europe 1792–1914* (2000).

For the twentieth century, recent studies include S.C. Tucker, *The Great War, 1914–18* (1998), A. Millett and W. Murray (eds), *Military Effectiveness in World War II* (1988), R. Reese, *The Soviet Military System* (2000), S. Sandler, *The Korean War: No Victors, No Vanquished* (1999), S.C. Tucker, *Vietnam* (1999) and S. Buckley, *Air*

Power in the Age of Total War (1999). The disparate elements of post-1945 military preparedness and conflict range from nuclear weaponry to guerrilla warfare. There have been very few successful attempts to put these together. Works worth reading include Martin van Creveld, *The Transformation of War* (New York, 1991) and *The Rise and Decline of the State* (1999). A critical view is provided by C.H. Gray, *Postmodern War: The New Politics of Conflict* (1997), and a different perspective by C.S. Gray, *Modern Strategy* (Oxford, 1999). A chilling portrayal of some recent conflicts can be found in A. Clayton, *Frontiersmen: Warfare in Africa Since 1950* (1999). On air power, see P.S. Meilinger (ed.), *The Paths of Heaven: The Evolution of Airpower Theory* (Maxwell Air Force Base, AL, 1997) and D.R. Mets, *The Air Campaign: John Warden and the Classical Airpower Theorists* (Maxwell Air Force Base, AL, 1998).

Other important recent studies on the here and now as well as the possible future include M.J. Mazarr et al., *The Military Technical Revolution: A Structural Framework* (Washington, DC, 1993), and K.P. Magyar and C.P. Danopoulos, *Prolonged Wars: A Post-Nuclear Challenge* (Maxwell Air Force Base, AL, 1994).

Index

305

Index

illiteracy, current problems, 273–4
imperialism, 5, 7–8, 51, 124, 128–9, 174, 190, 195, 200–1, 202, 203–4, 227, 280
Incas, 34, 105, 123, 125
India, Afghan campaigns in North, 15; British presence in, 20, 35, 37, 49, 113, 117, 121, 122, 123, 128, 129, 131, 132, 133–4, 135–6, 201, 225, 227; conflict with China (1962), 228, 267, 291; conflicts with Pakistan, 23, 67, 228, 230, 255, 260, 266, 290; end of British colonial rule, 22, 204, 225; French expansionist policy in, 181; military capability today, 253, 260, 265, 275; Mughal conquest of, 20, 34; use of cavalry in eighteenth century, 41; *see also* Marathas; Mysore
Indian Mutiny (1857–8), 74
Indian Ocean, 10–11, 49, 95, 105, 110, 113, 123
Indo-China, 40, 228, 243
Indonesia, 19, 33, 78, 81, 84, 86, 204
infantry, 18, 36, 50, 51–2, 52–3, 55, 62, 97, 102, 168, 242, 250, 296; in British Civil Wars, 148, 149; in World Wars, 210, 215, 217, 219–20, 241
information, and warfare, 239–40, 261, 285
insurrectionary struggles, 84, 226, 227, 228, 229–31, 246, 256, 264, 265
Iran, 23, 228, 230, 265, 266, 282, 284, 291, 293
Iraq, 33, 68, 69, 82, 204, 276; Gulf War, 228, 246, 256, 257, 261, 265, 274, 288; as ongoing threat, 250, 266, 281, 282, 285, 291, 293; suppression of Kurds and Marsh Arabs, 23, 81, 230, 265
Ireland, 19; 144, 145, 146, 150, 151, 169-70, 189; *see also* Northern Ireland
Irish, use of shock tactics, 42
Islam, 41, 202, 255, 263, 273, 282
Israel, 19, 69, 80, 215, 228, 229, 242, 253, 255, 259, 282, 283
Italy, 20, 69, 120, 189, 194, 204, 214, 218, 221, 222, 227, 247

Jackson, Andrew, 79, 101–2
Jacobites, 164–5, 167–9, 169, 170–1, 172, 190
Jamaica, 12, 85, 152, 227
James II, King of England and Scotland, 153, 169, 175
Japan and Japanese, 11, 21, 39, 48, 62, 79, 198, 199, 202, 212–13, 225; Aum Shinrikyo nerve gas attack, 265; navy and naval power, 43, 124, 197, 199, 259, 260; operations in China (1930s), 84, 204, 205, 206; Second World War, 16, 62, 204, 214, 215, 216, 217, 223, 240; use of atomic bomb against (1945), 248, 279
Java, 111, 121, 135
Jenkinson, Charles, 56, 65

Kashmir, 202, 255
Kashmiri secessionists, 23, 230, 255
Kazakhstan, European conquest of, 124
Kazan, Khanite of, Muscovy's conquest of, 119, 120, 130, 132

Keegan, John, 217–18, 275–6
Korea, 43, 124, 130; North, 250, 265, 266, 281
Korean War (1950-3), 76, 225, 227–8, 231, 249, 267, 291
Kosovo conflict, 16, 24, 32, 237, 245, 250, 252, 262–3, 265, 289
Kurds, suppression of, 23, 81, 203, 230, 265, 291
Kursk, Soviet defence of, 214, 219, 223

land, and conquest, 9–10, 145, 152
Latin America, 5, 31, 50, 74, 81, 83, 84, 85, 132, 203
Lebanon, 207, 255, 258, 291
Lee, General Robert E., 144, 158, 159, 160, 162
Leveller rising (1649), 151
Liberia, 22, 82, 83
Libya, 81, 82, 120, 258, 266, 276, 284
Lithuanians, 40, 206
Livonia, Russian conquest of, 99
local wars, of the British Civil Wars, 147–8; wider implications in current climate, 287
Louis XIV, King of France, 175, 176, 180, 183, 184, 185
Ludendorff, General Erich von, 214
Lynn, John, 14, 37–8, 39, 41, 44–5

MacArthur, General Douglas, 68, 79
McClellan, General George, 156, 158, 159
Madagascar, 9, 82, 121, 125, 206, 207, 209
Magyars, 40
Malacca, 9, 122, 258
Malta, naval attack on by Ottoman Turks, 10
Manchuria, 130, 204, 206, 246
Manchus, 20, 34, 36, 96, 98, 112, 129, 131
Maori, tactical sophistication, 17
Marathas, 50, 51, 73, 97, 125, 132, 135, 136, 174
Marlborough, John Churchill, 1st Duke of, 48, 99
Marx, Leo, 93
mercenaries, 33, 56, 70, 79, 113, 155
Mesopotamia, Turkish defeat of Western armies (1916), 13
Mexican War (1846–8), 155–6
Mexico, 85, 105, 111, 121
Middle East, 69, 81, 203, 227, 228, 229, 238, 253, 273, 274, 284, 289
military capability, 7, 12, 77, 96, 98–9, 103, 104–5, 124–5, 171, 196, 208, 238, 257–8
military coups, 67, 81–2
military history, approaches, 1–3, 13–14, 28–30, 86, 90, 95, 104–7, 112–13, 115, 218, 237, 252; role, 3, 24, 222–3, 297
military organizations, development 1850–2000, 62–86; development to 1850, 28–57; modern, 199, 240–2, 244, 253, 296–7
military revolution, 55; 92–3, 93-4, 95, 97, 99, 104-36, 278, 293; *see also* RMA
military service, attitudes, 29, 30–1, 243–4, 246, 247–8, 292–3; *see also* conscription; recruitment